Classic Lanterns

A Guide and Reference

Dennis Pearson

Revised and Expanded 2nd Edition

4880 Lower Valley Rd. Atglen, PA 19310 USA

Dedication

To the man who gave me my third lantern,
my Brother Don, the hardest working man I know.

Front Cover: Dietz Pioneer Street Lamp made between 1906 and 1944. Refer to plate 6.3 for more information.

Library of Congress Control Number: 2008922896

Designed by Bonnie M. Hensley
Typeset in Futura Hv BT/Times New Roman

ISBN: 978-0-7643-2876-3
Printed in China

Published by Schiffer Publishing Ltd.
4880 Lower Valley Road
Atglen, PA 19310
Phone: (610) 593-1777; Fax: (610) 593-2002
E-mail: Info@schifferbooks.com

Please visit our web site catalog at www.schifferbooks.com

Contents

Acknowledgments

I wish to express my sincere thanks to all the people who contributed information, advice, support, and the lanterns that made this book possible.

People who helped put this book together: Jeff Snyder, Nancy and Peter Schiffer.

People who gave me lanterns over the years, Frank L. Colwell, Juditha Colwell, Alice and Ted Knutson, Glenna Moore, Carl D. Pearson, Donald R. Pearson, Lorna J. Pearson, Laura and Frank Pelligrino, Anita, Gerald and Danny Puetz.

People who gave information and support: Gilbert Belcher, Rita Heacock, George Lawless, Robert Olsen, John Salvestrini, Dawn and Woody Kirkman, Phil Embury, George A. Vandercook, and Tom Vause.

People and organizations who loaned lanterns from their collections: Scott Schifer, Janet and Martin McGuire, Gilbert Belcher, and the Belmont Shores Model Railroad Club.

And thanks to the hundreds of antique dealers and workers in southern and central California for many pleasurable hours browsing for interesting and unusual artifacts of our industrial heritage.

Plate 0.1
Adlake Kero railroad style lantern marked BR, with an amber globe.
From the collection of Janet and Martin McGuire.

Introduction

This book is an effort to document the history, contribution and technological progress of lanterns in the United States. Much can be learned from the study of lanterns because the development, refinement, and production off shore parallels the evolution of many other industrial consumer products such as autos, televisions, VCRs, steel, and computers.

As with other consumer products, lanterns have their history, great inventors, design infancy, mass market maturity, art deco heyday, and decline. Lanterns still serve a practical purpose today in rural and urban situations alike. For frugal campers, a kerosene lantern is a practical substitute for more expensive gasoline, propane or electric lanterns. Also, lanterns serve today as decorative accents in theme parks and restaurants across the country. The film industry has long used kerosene lanterns, placed on sets to represent a rural or depressed living situation, not to mention the many films with Old West themes.

Kerosene lanterns helped to develop the great plains of the central United States before great hydroelectric power projects of the 1930s. Cities of the industrial East had gas and electric lights before 1900 but wide distribution of electric power in the dry Southwest was slow to develop. What the southwest had was kerosene from the oil fields of Texas and California. The farms of the midwest were likewise isolated, and relied heavily on lamps and lanterns until rural electrification.

This book concentrates on the classic American barn style lantern but touches on related lamps and lanterns for comparison. Hurricane lamps, railroad lanterns, modern and ancient lanterns are discussed in context as they relate to kerosene lanterns.

It would be impossible to document every lantern from every manufacturer in the U.S. and this book does not attempt to represent all aspects of lanterns and the variety of manufacturers, styles, and lantern uses. It describes the history, development, operation, styling, construction, and materials that will allow identification of the age and value of almost any kerosene lantern.

Each lantern manufacturer made continuous minor changes to their products to maintain their fickle hold on the market. While using this book to identify a particular lantern, bear in mind that there may be subtle differences between lanterns of the same make and model. Take, for example, the two Dietz D-Lites below. Note that they are the same lantern in every respect except for the fount capacity. The D-Lite with the larger fount was called "special" and cost just a few cents more. The difference is nothing more than a way to squeeze another nickel out of the consumer. These sales tactics are still common today.

We will see upstarts like the Defiance Lantern and Stamping Company, whose very name was a shot across the bow of the big lantern companies. We'll see exaggeration in the advertisements, companies jockeying for greater market share, and the acquisition of competitors until only a few companies remained.

The story of the oil and kerosene lantern enterprise also parallels the stories of other major U.S. industries; steel, railroad, oil, auto, textiles, on and on. The players are not as well known as Ford (cars), Rockerfeller (oil), Carnegie (steel), Getty (oil), Vanderbilt (railroads), or James Pierpoint Morgan (finance) but, their stories are just as interesting.

I hope you enjoy this stroll through American history from the flickering perspective of a petroleum flame.

Plate 0.2
Note the difference in fount size on the two 1919 Dietz D-Lites.

Foreword

There are many different facets to kerosene lantern collecting. There is the search for more perfect examples to fill a collection and the effort to collect lanterns from different countries and of different sizes or colors. Many people enjoy poking around antique stores, flea markets, and yard sales. For some, the goal is to own "one of everything;" others enjoy repairing and restoring lanterns to operating condition. Most collectors also look for paper records, ads, and catalogues for their own research and to compliment their collection.

Kerosene lantern collecting is not limited to antiques because new lanterns are being made, sold and used around the world. Many unusual and interesting lanterns appear in sportsman, decorating, and hardware supply catalogues regularly. In the United States today, the premier distributor of kerosene lanterns is V&O Sales, Inc. of Syracuse, New York. Tom Vause and Bob Olsen of V&O are former R. E. Dietz employees who now operate their own wholesale and retail sales corporation. They distribute V&O brand lanterns in a variety of sizes and colors.

Tom Vause had been with the company for thirty-five years and was the National Sales Manager when Dietz closed its doors in 1992. I asked Tom to add his thoughts for the readers:

Prior to coming to the Lantern Works, as it was known, I was employed by the Illinois Bell Telephone Company as a Service Representative. I was asked to join the Dietz Company by a fellow who I met while in the Army in the early fifties, and with whom I had kept in contact. This was my first close encounter with kerosene lanterns other than those I had seen in Western movies.

My first position was as a Territory Salesman in the middle western states, calling on all the old hardware and industrial supply houses that supplied lanterns for rural areas without power, Amish folks who do not use electricity, and for safety lighting around construction sites.

Even though Dietz had moved their lantern operation to Hong Kong in 1956, they were still producing a few replacement burners and highway torches in the Syracuse, N.Y. plant. Most machinery was still operated from what were known as main transmission lines in the ceiling. This power would run a multitude of machines driven by large leather belts. This did not last very long after the government formed OSHA, the Occupational Safety and Health Administration.

During Dietz's last sixty years, it moved into the heavy duty truck lighting business and spent most of the time and capital on this line, while the lantern business slowly diminished. The market for lanterns was strongest in the Middle East, Far East, and third world countries where there was little electrification. For this reason Dietz moved the entire lantern operation to Hong Kong.

When I first started working for Dietz, the average time of service per employee was twenty eight years. When they closed in 1992, it was down to six years. There were three generations of Dietzes working with the firm when I started in 1963, and only one when I was terminated in 1992.

In my estimation, there were two major reasons for the demise of the R. E. Dietz Co.: a long-lasting union strike which was still in effect when the company closed, and a lack of sound management by the executives of the company. It was a difficult time for those employees that had expected to finish their working careers at Dietz. Very few Dietz employees kept their jobs after the new company purchased the Dietz major truck lighting division. The balance either retired early, if they could, or searched out new jobs. The last Dietz family member to be involved in the company, Hugh Dietz, is presently selling investment portfolios, to the best of my knowledge.

I was one of the long term employees who was not old enough nor ready to retire. My superior at Dietz, Bob Olsen, and I had begun marketing Dietz lanterns in the late eighties, so we formed V&O Sales, Inc. and continued what we had begun. Today, the use of lanterns is an efficient lighting source and a nostalgic reminder of seemingly easier and less complicated times, which many fondly remember.

Tom Vause

Plate 0.3
Classic barn lanterns are still being made and sold. Refer to the Web Directory in Chapter 12.

Chapter 1

History of The Oil Lantern

The lantern must have been invented the night a brave cave woman first pulled a burning stick out of the community fire and use it to light her way to the prehistoric powder room. By our definition, a burning stick is a lamp with wick and fuel combined in a handy, one piece, package. This burning stick was about as far as lantern technology progress for the next million years. We can speculate that bundles of dry grass or dung attached to a stick made an effective torch. The British still refer to a flashlight as a "torch."

By the dawn of recorded history, a crude oil lamp had been in use for some time. The ancient Greeks, Romans, and Egyptians each made very similar lamps consisting of a pottery oil container with a hole for a wick (Plate 1.1).

The ancient Persians further refined the simple pottery lamp as well as making copper, bronze, and even gold ones (Plate 1.2). These lamps were still very simple affairs of a fount and a wick. The fuel was usually cheap and plentiful animal fat that must have smelled horrible. The lamps had no chimney so the flame was very yellow, the oily soot was awful and they didn't work well out doors.

Miners from pre-recorded history until the mid-19th century used oil fired lamps that changed only slightly in thousands of years. The Appalachian coal miners must have been proud that 6000 years of technological innovation had provided their lamps with a lid to keep the oil from spilling (Plate 1.3).

Plate 1.1
This Egyptian clay oil lamp is a reproduction of the type that was used around 300-600 AD (1.75 in. H, 4.0 in. W, 2.13 in. D)

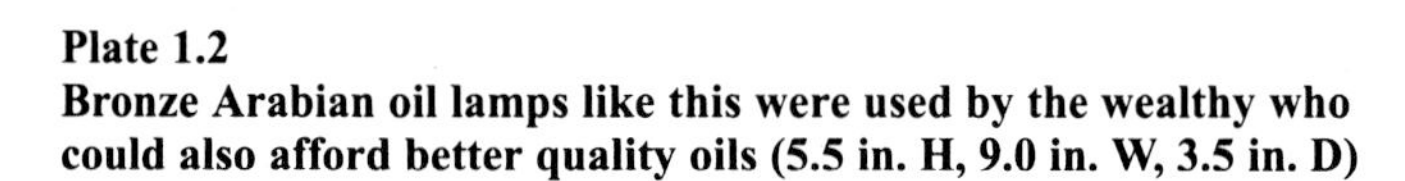

Plate 1.2
Bronze Arabian oil lamps like this were used by the wealthy who could also afford better quality oils (5.5 in. H, 9.0 in. W, 3.5 in. D)

Plate 1.3
This nineteenth century copper miner's lamp is marked:
STAR GRIER BROS.
PITTSBURGH, PA
(2.75 in. H, 1.65 in. W, 3.5 in. D)

Eventually, candles made of tallow (or bee's wax if you could afford it) became the fuel of choice since it didn't spoil and was easier to transport than unprocessed animal fat. Once the candle became a light source the candle lantern was right behind (Plate 1.4). The lantern of the mid-17th century was a candle in a tin and glass cabinet. Candles remained the most abundant source of illumination for the next 200 years.

With the improvements in ships and navigation there was a return to oil fired lanterns in the 18th and 19th centuries. Processed sperm whale oil was found to burn clean, smell decent, and not spoil as other animal fats did. Up to two thousand gallons of lamp oil could be rendered by boiling the blubber of one large sperm whale. Even so, whale oil was only used in the lamps of the wealthy. The working class still used tallow candles for indoor lighting.

Plate 1.4
This folding candle lantern is made of galvanized steel and isinglass (mica):
THE "STONEBRIDGE" FOLDING LANTERN
N.Y.C. MADE IN U.S.A. (10.25 in. H, 4.5 in. W, 5 in. D)

In the last half of the 19th century, the "Industrial Revolution" produced vast advancements in lighting technology. You will recognize the gasoline lantern that is still popular with campers to this day (Plate 1.5).

The electric arc light, gas light, batteries, incandescent bulbs, gasoline, coal oil and kerosene all made their appearance just over a century ago. The use of Calcium Carbide to generate an acetylene flame was popular for both early auto headlights and miner's lamps (Plate 1.6).

In the late 19th century, whale oil became more expensive as whales became harder to find. Coal oil, starting in the 1830s, was the first fuel inexpensive enough to replace the candle in the homes of the middle class. When Edwin L. Drake sank the first successful oil well in 1859, cheap and plentiful kerosene was made from the crude oil. Many men became wealthy by drilling, pumping, cracking, and selling kerosene the world over. John D. Rockefeller (1839-1937) founded Standard Oil and became the world's wealthiest person, not by selling the useless byproduct gasoline, but by selling kerosene for lamps and lanterns worldwide.

Plate 1.5
The Coleman® gasoline lantern became the standard camping lantern of the 20th century:
COLEMAN TRADE MARK REGD
The Sunshine of the Night
Made In The United States of America
10 56
(15.25 in. H, 7.5 in. W, 7.0 in. D)

Plate 1.6
Acetylene miner's lamps, left, GUY'S DROPPER, top, AUTOLITE with a can of Calcium Carbide. The GUY'S DROPPER on the right is marked:
PAT 5-2-12, 5-26-14, 9-19-16, 11-2-20, 2-10-25, 7-21-25
(4.25 in. H, 3.25 in. W, 3.75 in. D)

Lanterns of the 1850s and '60s come in two broad styles: the box lantern and the glass globe models. Improvements in manufacturing, transportation, and communications, during and after the Civil War, helped to bring the price of glass and tin down. The hand lantern was basically a lamp made of tin with a rain hood over the chimney. The chimney soon gave way to a more practical globe and the lantern began to have the appearance of a modern dead flame lantern (see Plate 1.7).

Plate 1.8
Early hot blast lantern made under a license granted by John Irwin: Steam Gauge & Lantern Company L W (14.0 in. H, 7.25 in. W, 5.75 in. D)

Plate 1.7
Civil War era dead flame oil lantern: (19.0 in. H, 8.0 in. D).

In the 1870s, the lantern changed again because of the patent awarded to the prolific inventor John Irwin on January 12, 1868 (patent number 73012). The patent described tubes that diverted some of the draft to blow on the flame, making it burn hotter and brighter. The early "tubular" lanterns, like the S. G. & L. Company's LW (Plate 1.8), were as simple as possible. Their construction was just round or square segments of tubing soldered together by hand. By the 1880s, lanterns had improved tubes of pressed steel strengthened by ridges and crimped together. Founts were still small and fill caps were even smaller. Construction was still mostly done by hand.

Technology and design took a quick leap forward in 1880 with the invention of the cold blast draft. Its design, used in the C. T. Ham No. 2 of Plate 1.9, produced an even brighter light by forcing even more oxygen into the flame. The problem with the hot blast technique was that spent air was being forced down to the burner. Spent air from the flame has much of its oxygen used and more carbon dioxide. The cold blast design used the rising warm air of the flame to create a draft that draws fresh air through the tubes down to the burner. Just like blowing on a fire, the flame burned brighter.

Plate 1.9
Early cold blast lantern: C. T. HAM MFG. CO. ROCHESTER NY USA No. 2 COLD BLAST (14.5 in. H, 7.0 in. W, 7.0 in. D)

The patents were tightly controlled but many independent stamping companies produced hot and cold blast lanterns without paying royalties. These stamping companies found themselves being sued by John Irwin and the companies that had paid for the rights to use his patents.

Steady improvements continued to be made through the teens and twenties including better globe lifts, various reflectors, larger fount caps, lenses, LOC-NOB globes, and much more.

By 1920, the construction of the kerosene lanterns was fully standardized. Only a handful of globe styles were made and these were interchangeable between lanterns of different manufacture. Those stamping companies that survived the depression had eliminated most hand operations to reduce cost to a minimum. Solder was kept to a minimum in the construction of lanterns. Steel replaced brass in burners, fount caps, and wherever else possible.

Since the technological development of lanterns was complete the next step in lantern evolution was an obvious one . . . styling. The bleakness of the economic depression fueled the free-form styling that is now known at Art Deco. Lantern designs from the remaining producers focused on the shape, style and even color of what had previously been a strictly utilitarian object. The Dietz Monarch, designed in 1936, (Plate 8.2) is an excellent example of deco lantern styling. The curved tubes with decorative crimping give the lantern a rounded, almost tear drop shape, popular at the time. The use of colors took a step forward when lanterns began appearing in metallic green and metallic blue.

A legislative upset eliminated a major lantern market in the last half of the twentieth century. Until 1955 the use of lanterns to mark street work, excavations, and public work projects had been steadily increasing. Less expensive substitutes for lanterns, called torches, were developed for contractor's use. The potential for starting fires is obvious and in 1955 the Federal Government passed a law that eliminated the used of fire as hazard warning device on public highway projects. Similar state laws soon followed.

The logical result of the lantern ban was the familiar, battery powered warning flasher in Plate 1.10.

Battery powered lights for railroad use had been growing in popularity through the 1920s, 30s, and 40s. Railroad companies continued to buy kerosene lanterns in decreasing numbers into the 1960s even though perfectly good battery lanterns were available (Plate 1.11). Kerosene lanterns continued to be used for private and municipal works but the fire hazard, loss due to theft, breakage, and the improvements in outdoor plastics and rechargeable batteries caused production of contractor's kerosene lanterns to halt in the mid 1960s.

This should be the end of the story, but kerosene lanterns continued to be sold in America and abroad. Foreign customers continued to buy lanterns from R. E. Dietz Limited, who moved production to Hong Kong in 1956. Various models were available in the US throughout the 1960s, '70s, and '80s (Plate 1.12). Two notable special editions were issued during this period: the bicentennial lanterns in 1976, and the Dietz 150th anniversary reissue of the Junior in 1989. Electrified versions of the old west street lantern were sold by the hundreds to Disneyland,

Plate 1.10
Battery powered warning flasher of the type still in use today: FLEX-O-LITE ST. LOUIS, MO NIGHTFLASHER (12.5 in. H, 7.5 in. W, 3.0 in. D)

Plate 1.11
Battery powered railroad conductor's lantern:Conger Lantern Co. of Portland Oregon (made from 1917 to 1989, 10.5 in. H, 6.5 in. W, 4.0 in. D)

Plate 1.12
R. E. Dietz advertisement from 1982 does not mention the seven other models that were also made in Hong Kong.

Disneyworld and individual landscapers. Campers purchased a few and emergencies caused by natural disasters sold few more.

In September, 1992, after 152 years of lamp, lantern, and automotive lighting production, the doors of the R. E. Dietz company were closed forever. Several reasons for the closure are noted. A crippling UAW strike halted deliveries of electric truck lights that were the mainstay of the Dietz production (Plate 1.13). The fifth generation of the Dietz family was beyond retirement age and the sixth was not interested in continuing the business, so the company was liquidated.

Although the Dietz company was closed, the production of Dietz lanterns continued thanks to two former Dietz employees, Tom Vause and Bob Olsen. Vause and Olsen secured the rights to import Dietz lanterns to the U. S. market. Their company, V & O Sales, has imported eight different Dietz models in several colors since 1993. They are: Air Pilot, No. 50 (Comet), D-Lite, Junior, Blizzard, Wizard, Monarch, and The Original.

Today a wide variety of new kerosene lamps, lanterns and parts are available. The companies Adlake of Elkhart, Indiana; V & O Sales of Syracuse, NY; and Lamplight Industries of Menomonee Falls, Wisconsin sell new lamps, lanterns, replacement parts and supplies. Nearly all lamp and lantern items are now made in The Peoples Republic of China. These imports are inexpensive and of high quality. For comparison sake, we note that in 1982, a solid brass lantern from the Dietz factory in Hong Kong sold in America for $85 US. Ten years and 30% inflation later a similar solid brass lantern from China sells for $65.

Today's kerosene lantern is as functional as ever for camping, decorating, romance, and natural disaster preparedness. Kerosene lanterns are less expensive to operate than battery powered camp lanterns and less expensive than gas lanterns to buy. Old lanterns compliment Contemporary, Southwestern, Early American, and Victorian decors. New lanterns are a decorator item that should be available in electrified form from home centers and lighting stores. Unfortunately, today lanterns are available in only a few hardware stores, catalogues and gift shops.

Plate 1.13
Dietz amber clearance light as seen on thousands of trucks world wide.

Chapter 2

Operation of The Kerosene Lantern

The lanterns in this book use the capillary action of a wick to draw the fuel to the flame. (The meaning of "wick" is capillary action.) By this definition a lantern is just another form of candle. A candle is an absorbent wick surrounded by solidified fuel (i.e., paraffin wax). As the flame heats the wax it liquefies and the capillary action of the wick carries the liquid fuel up to the flame. The heat of the flame turns the liquid fuel into a flammable hydrocarbon vapor that combines with nearby oxygen and is consumed.

Gasoline lanterns of the Coleman® type (plate 1.4), although similar in function, operate on a different principle. A Coleman® lantern uses a pressurized tank to force liquid gas into a tube near the hot mantle. The heat from the mantle "boils" the liquid gasoline and makes a gas vapor that is consumed in the mantle.

In a properly functioning kerosene lantern it is the fuel that burns, not the wick. It's even more complicated than that because liquid kerosene will not burn. The liquid fuel (oil or kerosene) must be vaporized by the heat of the nearby flame and mixed with oxygen before it will burn.

Adding more oxygen to a flame makes it burn hotter and brighter. The light from an open flame, as in a candle, can not exceed one candle power because it has only oxygen from the surrounding air to draw from. If enough oxygen is added to a flame it can burn through steel. In fact that is how an acetylene cutting torch can cut through steel.

Burning more fuel (i.e., making the wick larger) without adding more oxygen just produces incomplete combustion. A smoky yellow flame is the result of incomplete combustion. Soot, also called lamp black, is almost pure carbon from fuel not converted to carbon-dioxide (burned) due to a lack of oxygen. The operation of a forge, where air is forced into a flame under pressure to raise the temperature, has been understood for centuries. Early inventors soon discovered that supplying oxygen to the flame is the key to a brighter light.

The first lamps were just an open flame like a smoky camp fire. Once the advantage of the chimney was understood, a glass chimney was added to candles and lamps. Heat from the flame creates a draft in the chimney so more air is pulled in below the flame. The extra oxygen allows more fuel to burn and makes a brighter flame. A candle with a chimney can give more than one candlepower of light. By the 15th century the high cost of glass was the only thing preventing everyone from having a brighter lantern.

As oil lanterns developed, several different methods of getting oxygen to the flame were tried. The glass chimney or globe is the simplest and most obvious improvement and really defines what we call a lantern today. Some of the earliest lanterns are a lamp inside a box. The surrounding box is an attempt to weather proof the lamp for outdoor use. The lanterns of Plate 6.1 are this kind of lantern.

Tubular Lanterns

The term "tubular" is a reference to any lantern with a tube or tubes used to carry air to the burner for better combustion. The hot blast and cold blast are the best examples. A hot blast lantern has a tube or tubes that carry used exhaust gas from the chimney to the burner. A cold blast design routes fresh outside air to the flame. Before tubular lanterns there were only "dead flame" types that had a chimney but no forced air system built in.

The need for a brighter lantern led to the development of the hot blast tubular technology. In 1868, John Irwin was awarded the patent for the tubular lantern. In a tubular lantern the draft is provided by the rising warm air from the flame being returned through the warm air tube(s) to blow on the flame. Plate 2.1 shows the crown of a common hot blast lantern and Plate 2.2 is a cut-away diagram showing the air flow in of a typical hot blast lan-

Plate 2.1
Tube arrangement of a typical hot blast lantern.

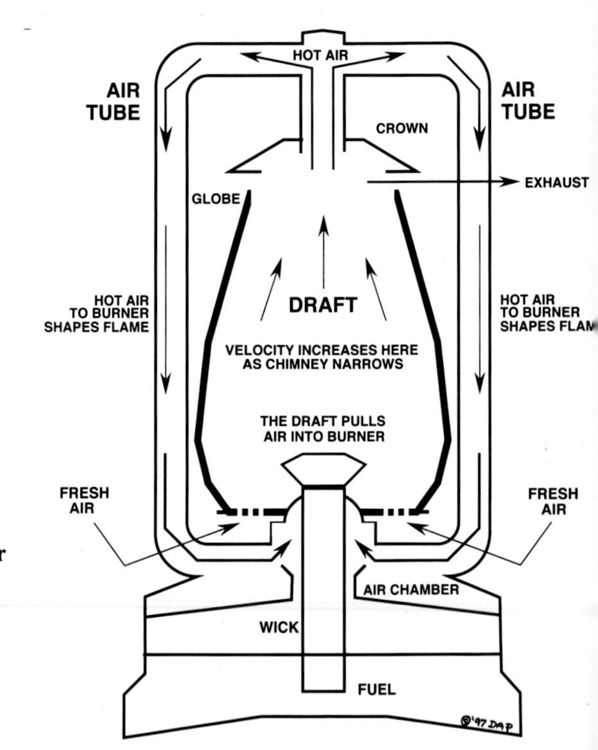

Plate 2.2
Cut-away diagram of a hot blast type tubular lantern, showing exhaust gasses being forced into the burner.

tern. Both hot blast and cold blast lanterns are tubular. Lamps and dead flame lanterns are not even remotely tubular.

In due time the tubular design was improved upon and the cold blast lantern was developed.

Cold Blast Lanterns

Cold blast (or C.B.) is often included in the lantern name or description because it was a major innovation. The reference is to the fresh air vent and globe retaining assembly at the top of this type of lantern. Plate 2.3 shows the chimney assembly of a typical cold blast lantern. The advantage of the cold blast system is a brighter light without going to a larger wick.

In the cold blast design, more fresh air is used to force-feed the flame. Plate 2.4 is a functional diagram with arrows indicating the air flow of a cold blast lantern.

Plate 2.3
The cold blast chimney pre-warms the fresh ambient air before it gets drawn into the burner.

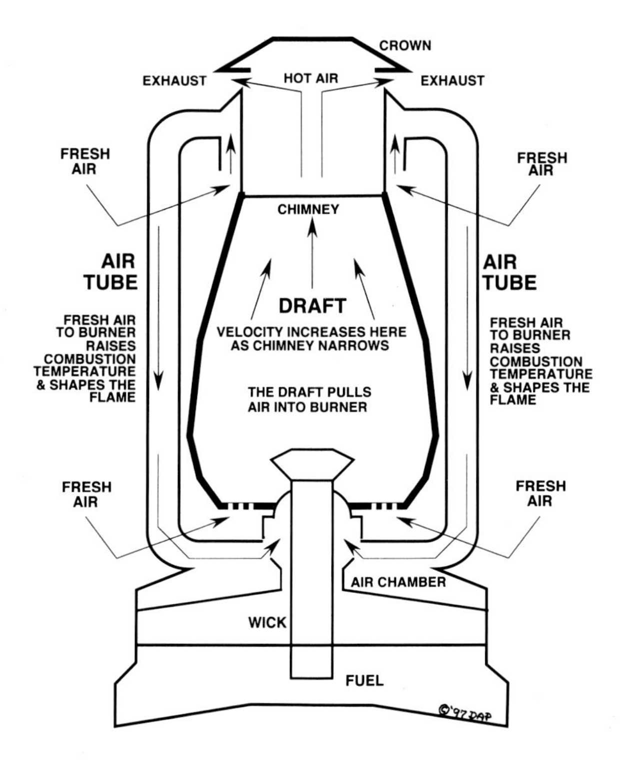

Plate 2.4
Diagram of a cold blast lantern showing the use of fresh air to supply more oxygen to the flame.

Parts of a Lantern

Throughout this book the features, innovations, similarities and differences of lanterns will be described by referring to the bits and pieces that make up lanterns. Some parts are common to all lanterns like, wicks, burners and globes. Sometimes the same parts may have different names like, the fount of a railroad style lantern is called a pot. Plate 2.5 identifies the important parts of a generic lantern. Refer to the glossary for descriptions of other lantern parts.

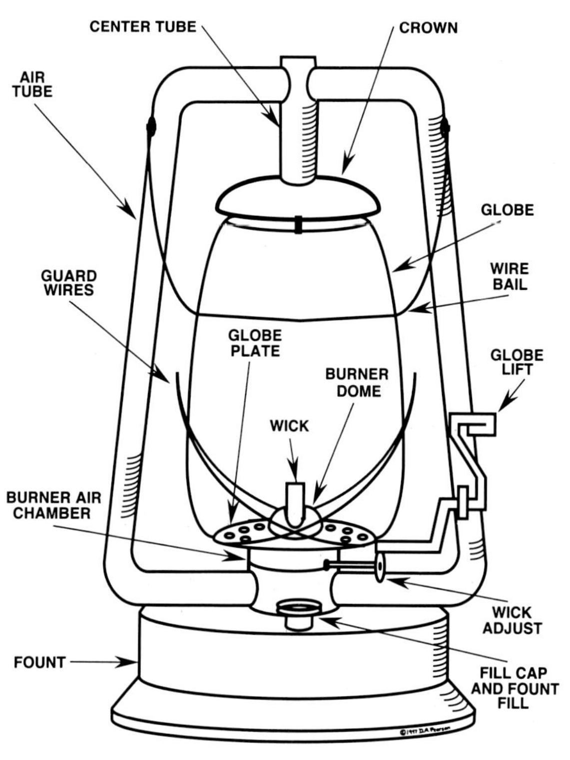

Plate 2.5
Some important parts of the generic kerosene lantern.

Lantern Terms

Most of these terms are no longer in common use yet, less than 100 years ago, they were well understood by anyone going to the hardware store to buy a new lantern. Some of the more common terms are described here and the rest can be found in the Glossary.

Globes

A globe is a tube shaped glass enclosure which replaced the flat plate of glass used on the earliest, square lanterns. The number zero (hereafter No. 0) globe is the standard for kerosene barn lanterns. Other globes include the "short" style used on the D-Lite, the smaller Junior, the huge No. 3 for street lights, the No. 39 railroad globe, the little wizard, and other special types.

There were variations of each globe type as well. The standard No. 0 globe was improved upon by Dietz when a lip was applied to the top ridge to help hold it in place. Patented in 1914,

Dietz globes were further improved with the addition of the LOC-NOB ears. Nearly all types of globes came in clear and ruby red. The ruby globe was used for signals and safety lights. Most of the globes in Plate 2.6 are ruby color. Blue, yellow, and green glass globes were used for special purposes.

Plate 2.6
A dozen of the most popular globe styles from tallest to shortest are: No. 0, LOC-NOB, junior, No. 39, Air Pilot, short, Vesta, wizard, kero, fresnel, original, and comet.

Burners and Wicks

The wick is made of absorbent material, usually cotton, that moves the fuel, by capillary action, from the fount to the flame. Wicks can be made of other materials with felt coming in a distant second. Lantern wicks come in many sizes from 0.25 inch to 2 inches (0.64 to 5.1 cm). Some of the more common lantern burners are the No. 0 burner that uses a 0.38 inch (1.0 cm) wick, No. 1 burner that uses a 0.625 inch (1.6 cm) wick, the No. 2 burner uses a 0.875 inch (2.2 cm) wick, and the No. 3 burner uses a 1.5 inch (3.8 cm) wick.

Do not confuse the burner number with the globe or lantern numbers. Lanterns were called No. 0, because they used the No. 0 globe, while other lanterns were called No. 2 for the burner they used. The common No. 0 globe was mixed and matched with No. 0, 1, and 2 burners. The only way to reliably determine the burner type is to measure the wick.

Wicks come in smaller sizes for novelty lamps and larger sizes for street lamps and torches. Unusual wicks include a tiny 0.125 inch (3.2 mm) round, a 0.75 inch (1.9 cm) round for torches, and a duplex (double wick) streetlight burner using two 1.5 inch (3.8 cm) flat wicks.

Plate 2.7
The shape of the dome opening controls the air flow and the shape of the flame.

All tubular lanterns have a burner dome that is matched to the burner. The shape of the opening in the dome controls air velocity and flame shape. The best flame shape is a broad arc pressed very thin by the force of the draft. Some domes are a separate part as in Plate 2.7, some domes are attached to the burner, and sometimes the dome is attached to the globe plate as in Plate 2.8. When the dome is attached to the globe plate, it is called a rising cone burner.

Plate 2.8
In rising cone burners, the dome is part of the globe plate so access for lighting is a little easier.

The size and type of burner is matched with the fuel type, globe size, and draft requirements. In the early days of lanterns many burners were tried but only a few survived to modern times.

The railroads were the first big industry to operate outdoors at night and were the driving force behind lantern development. Railroads preferred to use a lantern fuel that was cheap and readily available. Each railroad had their own mixture called signal oil which usually included lard oil.

The task for lantern companies was the development of lanterns to burn the signal oil efficiently. Kerosene had been the public's fuel of choice for decades but it wasn't until World War One, when the use of lard oil was curtailed, that railroads began using kerosene in earnest. The oil pots of plate 2.9 show two signal oil burners and one type of kerosene burner.

Globe Lifts

The earliest dead flame lanterns have non-removable globes fixed into the frame and the pot is removable much like the later railroad style lanterns. When the tubular lantern was invented a way had to be found to raise the globe so some early lanterns have a grab on the globe plate. A finger hook on the crown above the globe is called a square lift, shown in Plate 2.10. This arrangement proved problematic as the heat concentration is the highest at this point and the finger pull is useless after the lantern has been lit. Lighting a cold lantern is no problem but woe to the finger that has to lift the globe to blow out the flame.

Plate 2.9
Lantern pots with signal oil burners top and right compared to a Convex kerosene burner on the left.

The best idea was the side lift lever that "cranks" the globe plate up. The side lift comes in several variations including the outside lift (plate 2.11), rear (plate 2.12) and inside lift (plate 2.13). They all work about the same but the inside version seems to have won the sales race.

There are some unusual lifts to be found. The Dietz "Original" in Plate 8.20 has a left side lift and Prisco uses a formed wire (Plates 7.16, 7.17). The lift that takes the prize for most complicated is the Hurwood Aladdin (plate 7.8). The Aladdin's lift has a cam that raises the chimney off the globe then swings the globe completely out of the lantern. It's truly innovative but the idea never showed up on any other lantern.

Plate 2.11
This outside lift was quickly replaced by the better protected inside lift.

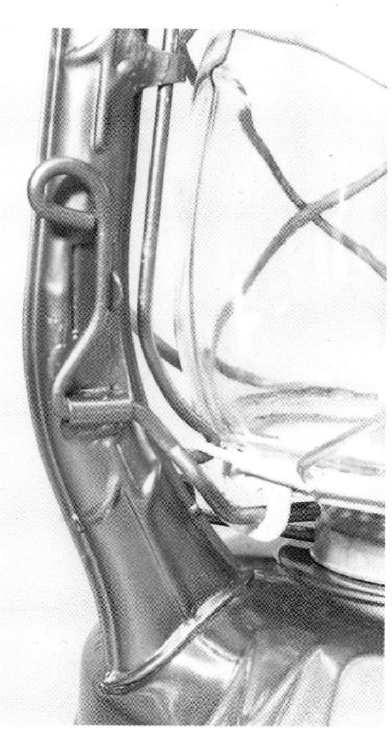

Plate 2.12
The rear lift is used on one of the most popular hot blast lanterns ever, the Dietz Monarch.

Plate 2.10
The top or square lift can give your finger a nasty burn when it comes time to blow out the flame.

Plate 2.13
The inside lift is the most popular and is in use on every cold blast lantern made.

Chapter 3

How to Use and Care For Your Lanterns

Cleaning

You just came up from your basement with that brass bottom Rayo left by the former owner. What is the best way to tackle it? First it must be cleaned to establish its condition and heritage (identify it). The best way to start is to give the lantern a soapy bath. Remove the globe, burner and fill cap. Use pliers if necessary but don't force them. If they won't come easily we will get them off later. I use a spray window cleaner on the globe and set it aside. Give the lantern a good shake to get out any sand, spiders, rocks, wasp nests, and lost wicks. The quickest way to recover a lost wick is to turn the lantern up-side-down and use needle nose pliers as shown in Plate 3.1.

Plate 3.1
Needle nose pliers are the best way to retrieve a lost wick.

Plate 3.2
Drop a small chain in and shake it around to knock lose any rust inside the fount.

Soak the lantern in a sink full of soapy water. Dish soap works fine, and if I don't care about the paint on the lantern, I use some ammonia. A good method for cleaning the inside of the fount is to drop in a small chain or some BBs and shake them around (Plate 3.2). It does a good job of loosening rust and stuck on crud. If you're in no hurry, put the parts in the dish washer with the next load of dishes. Why not? If it isn't rusted by now the water isn't going to hurt it. An old tooth brush is handy for getting soot from the crevices. Rinse the lantern in hot water and set it so the fount can drain. Examine the wick and if it is rotten, toss it. New wicks are available from your local hardware store.

Your Notebook

When all parts are dry, copy the maker, name, patents and other markings in your lantern notebook. Don't forget the marks on the burner and globe. Keeping a log of any collectible is a good idea not just for insurance or historic reasons but for future reference of forgotten details. My notebook is where I keep track of what I paid, good antique shops, clippings, articles, and all my receipts. Your lantern notebook is the one place to keep everything about your collection. I started with a 5 inch by 8 inch (12.7 by 20.3 cm) loose leaf notebook (Plate 3.3) which later proved to be too small. I now recommend a standard 3 ring binder for 8.5 by 11 inch (21.6 by 27.9 cm) sheets. I also record the unique features and details of construction. There should be a place for a photo and your sale price. Chapters 6 through 9 of this book are taken from my loose leaf notebook.

Plate 3.3
Research and record keeping may sound like a chore but, you'll be glad you did it.

Dealing with Fount Pin Holes

Study the fount bottom for signs of pin holes. Use fine sandpaper to remove any rust from the bottom. Unless the fount had water inside, the sides are usually better than the bottom. If pin holes are found, a drop of Super Glue™ will seal them. For more serious fount damage, refer to the restoration information in Chapter 10.

Care and Operation

New kerosene lanterns are available from better hardware stores, gift catalogues, and direct. Contrary to the way they are portrayed in the movies, modern lanterns (after 1912) are safe and easy to use. The fuel is barely flammable, the flame puts itself out if the lantern tips over, and the oil cannot spill. Each new lantern comes with printed safety warnings and instructions that should not be ignored.

Fuels

Reassemble the lantern and fill it with a good quality kerosene.

WARNING: *Use only approved lamp oil, paraffin, or kerosene. Do not use Coleman fuel, white gas, or gasoline.*

Both Coleman fuel and white gas are highly volatile unleaded gasoline. The use of gasoline in any wick type lamp or lantern would probably cause an explosion. Lamp oil and kerosene are more like a light lubricating oil than an explosive fuel, hence the safety advantage of kerosene lanterns.

Coal oil was developed in the 1830s and it is a term we don't hear much anymore, yet we still use coal oil all over the place. Paraffin and alkane are other names for coal oil. Paraffin is a semi-solid or oily distillate of wood, coal, petroleum, or shale that is a complex mixture of hydrocarbons and is used in candles, coatings, rubber compounding, pharmaceuticals, cosmetics, and lanterns. Pure liquid paraffin can be purchased wherever lamps and lanterns are sold.

The distillate of crude oil, named kerosene, can be used as a penetrating oil, clock lubricant, natural insecticide, or jet fuel. Its use as a jet fuel might imply kerosene is highly explosive but let me assure you, an explosion is the last thing you want in a jet engine. Kerosene becomes flammable when it is in a gas or vapor state by being either atomized or heated. In a jet engine the fuel is both heated and atomized as the injectors spray fuel into the hot combustion chamber under very high pressure.

I don't recommend colored or scented oils in kerosene lanterns because of the residue they leave behind. When the kerosene evaporates the color dyes and scent residue can gum up the wick and leave a mess behind. Even some so called "pure" kerosene can leave a sticky residue after 20 years but there is no way to tell in advance. I judge kerosene by its smell. Regardless of the label on the can, I won't buy kerosene unless I smell it first. There are small bottles of lamp oil and paraffin that will work in a lantern just fine. Lantern fuel can be purchased in bulk or in any size from 1 pint to 5 gallons. The price varies so it's wise to shop around.

Plate 3.4
Fuel comes in a variety of sizes but always smell before you buy.

Wick Replacement

The wick not only carries the oil to the burner but also strains out dust, dirt, and rust. I would suggest that if an old lantern is not burning properly, either the air tubes are clogged with spider nests or the wick is clogged with dirt and rust. The nests can be cleaned out with wire and new wicks are available at better hardware stores. In my book, any store with wicks is a better store.

There is an easy way and a hard way to change a wick. Unfortunately the easy way is also the messy way. Don't try to thread a dry wick through the burner as it will only cause aggravation. Wet the wick with kerosene and let its lubricating properties help slip the wick through.

Wick Trimming

Lantern brightness and efficiency depends on the size, shape, and color of the flame. The flame shape is determined by the fuel supply, oxygen supply (draft), and the wick shape. Unevenness or fuzz on the wick causes the flame to burn unevenly and brightness is reduced. Through use, a wick will eventually char and this build-up of carbon reduces fuel flow and causes an uneven flame. Efficiency is why most old or new wicks require trimming.

The correct way to trim a wick is to cut it straight across using sharp scissors. There should be no fuzz and the edges should be square. This is the wick shape the lantern was designed for and the shape that will burn the brightest.

Contrary to logic, turning up the wick does not make more light in the long run. If the wick is set too high, more fuel is available, but the oxygen supply remains the same. This causes incomplete combustion and excess soot. The soot coats the inside of the globe until the light is completely blocked. A high wick will also char faster since the areas not needed for combustion are quickly burned.

If the draft is good and the wick is clean, why do some lanterns give poor light? It is possible that a poorly designed lantern might have gotten into production, but it is not likely. Lantern technology was well understood and the buyers were careful shoppers. Yet we still find lanterns, especially dead flame types, that do not give much light. This problem can usually be traced to mismatched replacement parts or the wrong fuel. Each lantern has a number of vent holes in the globe plate, in the chimney, the bell and elsewhere, that are matched for the burner, draft, and fuel type. Burners and globes are often missing or changed, and many early lanterns are not made to burn kerosene.

As an example, the tall No. 39 globe railroad style lanterns were specifically designed to burn signal oil or lard oil. Lard oil is made from rendered pig fat. Signal oil might contain up to 50% lard oil that is much thicker and less volatile than kerosene. The volume of the globe and the amount of fresh air allowed through the little holes are carefully balanced for the thicker oil's burning characteristics. Putting more volatile kerosene in a tall globe lantern means the wick must be turned down until it is barely visible inside the burner. The result is a tiny, useless flame. I prefer to keep my lanterns cholesterol free so I use olive oil in my tall globe lanterns. Olive oil is just as great a lamp fuel today as it was 5,000 years ago in the Mediterranean. The wicks can be turned up and the original performance is restored.

Plate 3.5
Wood match sticks make it easier to light a lantern.

Lighting Your Lantern

The wick should still be at the correct height from the last use. Lift the globe, light the wick, and lower the globe. To avoid that sharp, sickening snap that is heard when a cold globe cracks, turn the flame down. On most burners, turn the wick adjust counterclockwise to lower the flame. Keep the flame low for about 15 minutes or until the globe warms up. The flame may grow as the lantern warms and the draft begins to take effect. For maximum brightness, readjust the warm lantern to just under the point where it begins to smoke.

Decorating with Lanterns

The common kerosene hand lantern would seem suited only for hanging in the basement waiting for the next power failure. However, with the wide variety of unusual sizes, colors, and styles, the lantern can be put on prominent display all over the house. An elegant dining room table setting calls for candle sticks, but a pair of polished brass lanterns provides a novel twist on the theme (Plate 3.6). A wide variety of scented oils are available and a lantern never drips wax.

Kerosene lanterns are a natural for today's popular Country and Southwest decor. A country kitchen feeling, with its copper pots, plaid curtains, and sunflowers, would be enhanced by a sky blue or bright yellow lantern hanging near the stove. The Southwest decor is an informal mix of Mexican and Western flavors. What could be more Western than a rusty old barn lantern?

A logical location for a lantern display is on the fireplace mantle in the den. An exhibit of lantern history using three or four periods of Dietz Blizzards or a matched collection of sizes would make a striking mantle display. Boat lanterns are a natural around the bar, especially with other nautical accent pieces.

A street lantern would be a useful accent on a pole in a front or back yard. Electrified, the lantern can provide useful outdoor light or, with a kerosene burner, it will cast a more romantic glow. Hand lanterns can be hung from the porch or patio as decoration, for light, or filled with citronella oil to chase away the bugs.

Polished brass lanterns are the most elegant but don't overlook a lantern freshly painted in a contemporary accent color. Paint and lanterns mix well. Feel free to paint any lantern any color you want. Paint protects the metal from a lantern's number one enemy, rust. Old paint is easily removed professionally or using an environmentally safe paint stripper.

Lanterns are often used as decoration in bars, restaurants, and amusement parks. Lanterns hearken back to a simpler time when the pace of life was slower and more relaxed. Disney World, Six Flags, Universal Studios and the other parks have impressive collections of new and old lanterns prominently displayed. The best part is we know which lanterns are the good ones and they don't.

Plate 3.6
Brass lanterns fit well into a formal decor.

Plate 3.7
A table display using a lamp and lantern together.

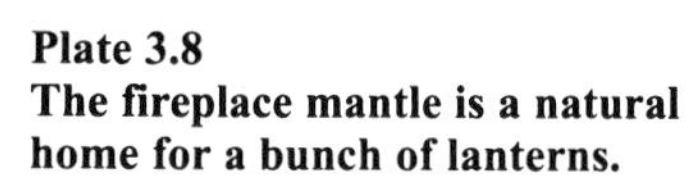

Plate 3.8
The fireplace mantle is a natural home for a bunch of lanterns.

Plate 3.9
A brass reproduction of a bulkhead lantern is used as a decorative accent.

Plate 3.10
Two 1936 Dietz Monarchs being used on a porch to chase away the summer bugs.

Plate 3.11
Lanterns come in a variety of colors or they can be painted to fit your needs. Here three red lanterns with ruby globes compliment each other and the room.

Lanterniana

Here is a word that has never been seen in print before. Lanterniana is objects, paper, and other material related to lanterns and lantern collecting. The image of the kerosene lantern has been used in movies, advertising, print, products, and paintings for 150 years. Lantern companies and distributors published catalogues that can be collected in original or reprint form. On the cover of the Saturday Evening Post of September 25, 1954, the American artist, Norman Rockwell, included a silver Dietz Vesta in his poignant painting, "Breaking Home Ties." A reprint of this Norman Rockwell painting can be the start of your lanterniana collection. The shape of the lantern has been use to make objects that will compliment any lantern collection as shown in plate 3.12.

Electrification Guide

Kerosene lamps and lanterns can be easily converted to operate on electric house current. Antique kerosene table lamps can use an electric conversion kit that replaces the burner. It's a good way to make the conversion since it does no permanent damage and the lamp can be easily converted back to oil later.

Although no such kit exists for lanterns, a plastic light bulb socket will fit in place of a standard barn lantern burner. A ceramic light socket is larger and will not fit. Try to avoid making holes in the fount or tubes as this reduces the lantern's value. I used an electrified lantern for a porch light for many years then put the burner back in with no harm done.

An unusual table lamp can be made from a large barn or railroad lantern using components from your local home center. To finish the lamp off, add a stable base of polished wood and a lamp shade. It is important to remember, never make holes in a kerosene lantern. A single tiny hole reduces the value significantly. A fetching night light or hanging accent can be made by the application of dried or artificial foliage as shown in Plate 3.15.

I have heard of lanterns being made into interesting bird houses by removing (and safely storing) the globe, and building a bird house to fit. A lantern that is hanging out in the weather should have a protective coat of paint every five years or so.

Plate 3.12
Examples of lantern related items include, from left to right, a miniature lamp, pencil sharpener, ancient lamp, Avon candy holder, key chain, and a Mastercrafters clock.

Plate 3.13
This rare lantern was all but destroyed by careless electrification.

Plate 3.15
Perhaps its dignity is lost but, an electric cord, plastic light socket and a seven and one half (7.5) Watt bulb, converts this Monarch to electric power without destroying its historic value.

Plate 3.14
A plastic light socket can be applied to the globe plate with hot melt glue.

Chapter 4

Kerosene Lantern Dating

Patent Dates and Numbers

A fairly accurate method for determining the approximate age of any lantern is by interpreting the patent dates. Before 1946, manufacturers often put patent dates on their products to help enforce their patent rights. Dietz, being the largest manufacturer, had the most patents to protect.

Look for patent dates stamped into the front of the right draft tube (Plate 4.1), into the top center draft tube (Plate 4.2), or on the draft chamber (Plate 4.3). The front of the lantern is the side with the fount fill cap.

Plate 4.1
Look for patent dates or numbers on the right air tube.

Plate 4.3
Prisco, Ham and other manufacturers put patents on the burner chamber.

Patent dates can also be found on the bottom, on the crown, or lightly stamped just about any place. Careful inspection is usually needed to find them.

The very last date tells us the lantern was made after that date, but not how long after. Some folks use the seventeen year rule. Since the patent is in force for only seventeen years, the manufacturer would have no reason to stamp a date beyond this. Without other evidence this may give a 20 year window of age but I have found exceptions to this rule.

The dates and numbers are often difficult to read but there is a trick I borrowed from the silver collectors. Use a candle to smoke up the numbers then lift the soot with a strip of transparent tape. Stick the tape to a piece of paper for a high contrast and permanent record.

Do not be fooled by the 1914 or 1923 dates on a globe (or any other date on a globe). Tubular globes are breakable and interchangeable from the earliest lanterns. Since the globe is most likely not the original, the printing on a globe is not an accurate indicator of the maker, or age. Do not fear. The information in this book will help identify most any lantern's age and the correct globe to use.

Plate 4.2
Patent dates may be found on the center tube.

Patent Numbers

Some manufacturers, including C. T. Ham, Embury and Dietz, used patent numbers instead of dates. Table 4.1 lists the approximate year for most patent numbers.

Table 4.1 Patent Numbers / Dates

for patent number	approx-imate year	for patent number	approx-imate year	for patent number	approx-imate year	for patent number	approx imate year
22477	1859	291016	1884	*910512	1909	1941449	1934
26642	1860	310168	1885	945345	1910	1985878	1935
31005	1861	*333494	1886	980178	1911	2026516	1936
34045	1862	355291	1887	1013095	1912	2066309	1937
37266	1863	375720	1888	1049326	1913	2104004	1938
41047	1864	395305	1889	1083267	1914	2142080	1939
45686	1865	418665	1890	1123212	1915	2185170	1940
51784	1866	443987	1891	1166419	1916	2227418	1941
60685	1867	466315	1892	1210389	1917	2268540	1942
72959	1868	488976	1893	1251458	1918	2307007	1943
85503	1869	511744	1894	1290027	1919	2338081	1944
98460	1870	531619	1895	1329352	1920	2366154	1945
110617	1871	552502	1896	1364063	1921	2391856	1946
122304	1872	574369	1897	1401948	1922	2413675	1947
134504	1873	596467	1898	1440362	1923	2433824	1948
146120	1874	616871	1899	1521590	1924	2457797	1949
158350	1875	640167	1900	1568040	1925	2492944	1950
171641	1876	664827	1901	1612790	1926	2536016	1951
185813	1877	690385	1902	1654521	1927	2580379	1952
198733	1878	717521	1903	1696897	1928	2624046	1953
211078	1879	748567	1904	*1719539	1929	2664562	1954
223210	1880	778834	1905	1742181	1930	2698439	1955
240373	1881	808618	1906	1787424	1931	2728913	1956
254836	1882	839799	1907	1839190	1932	2775762	1957
269820	1883	875679	1908	1892663	1933	2818567	1958

* estimated

Construction Details

The construction of lanterns changed to reflect the improved materials and machine tools available. The goal was to reduce costs, mostly by eliminating hand operations, and improve performance. The designs also changed as the use of the lantern changed from 19th century trips to the outhouse to 20th century street maintenance. Usage was a major factor driving design changes like the increase in fuel capacity over the years. Using the three generations of Dietz Monarch as an example, the fount capacity of the 1900 Monarch is 12 ounces. By 1912 the fount capacity is 16 ounces and in 1939 the capacity peaked at 18 ounces. The 1936 Monarch, with a ruby globe, was one of the most popular lanterns for utilities and municipalities nation wide. When filled to capacity with kerosene, the Monarch could warn motorists of a construction pit for up to 24 hours.

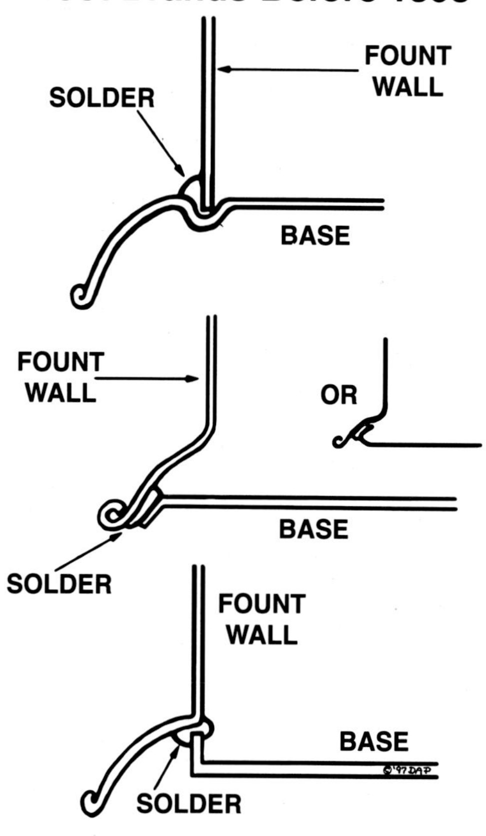

All Brands After 1900
crimped with no solder used

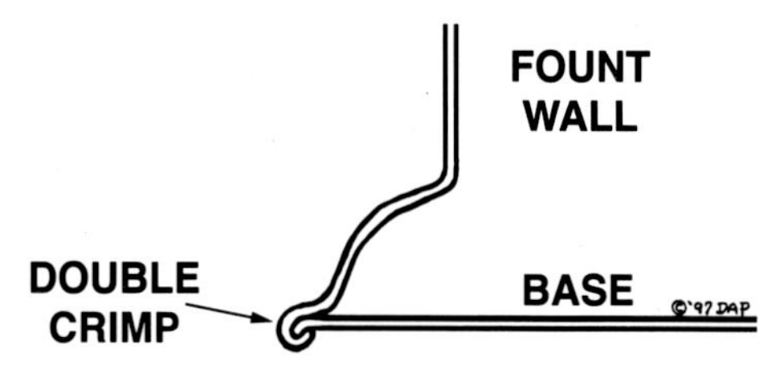

Plate 4.4
Before 1893 most founts were soldered by hand but by 1900 most lantern makers had changed to a crimp seam.

Fount Tubes

The reduction of hand assembly steps can be seen in the construction of the fount and air tubes. The early lanterns of chapter 6 were completely hand assembled using solder. As lantern designs improved, more and more seams were folded and crimped as typified by the lanterns of chapter 7. By 1912 only a few parts were attached with solder. As the country was recovering from the Great Depression, an investment in new tooling produced lanterns with nearly all crimped joints as shown in the lanterns of chapter 8.

Fount Cap Styles

The fount fill cap also changed through the years. Plate 4.5 shows a variety of caps from the small, one piece brass cap typical of pre-World War One lanterns to the present styles. In the early days fuel was expensive so the founts were small and the fuel fill hole was small. As the founts became bigger it made sense to fill them faster. All caps should have a red fiber seal. Other variations include a cap with the threads on the inside, hinged caps, and large caps made from one piece of brass, steel, aluminum, or plastic. Look for the small brass cap as an indicator of lanterns made before 1912.

Plate 4.5
Fill caps: Top row, pre 1912 Dietz, Ham, S. G. & L., second row, Prisco, early Embury, late Embury, third row, 1912 Dietz, 1936-1956 Dietz with full text, imported Dietz, bottom row, plastic Wheeling, Firehand (restrained).

Globe Styles

The invention of the glass globe marked a significant advance in lantern technology. Before the globe, flat plates of glass made up the sides of a square lantern. The earliest globes were mouth blown and no two were identical. Soon machines blew the glass into steel molds but there were no standard sizes. Eventually the majority of barn lanterns used the 6.75 inch (17.1 cm) tall No. 0 tubular globe shown in Plate 4.6. Through the years the shape of the globe changed but the dimensions stayed about the same. The fluted or pear shaped globe on the right was common before 1914. The globe on the left is a Dietz "LOC-NOB" that is dated 3-10-14. The early pear shape globe is a good indicator of an early lantern. Unmarked globes were used by smaller lantern makers and sold as replacements.

Collectors should seek out the early, pear shaped globes for use on early Dietz and other makes of lanterns.

What the heck is a LOC-NOB you ask? Notice the "LOC-NOB" globe in Plate 4.7 has bumps cast on each side of the glass about half way up. These are the "lock knobs." The globe guard on most Dietz lanterns made after 1914 were criss-cross loops of wire that used these knobs to lock the globe in place. Previously, when changing the wick or cleaning the burner, the globe was free to roll off the table and onto the floor. The LOC-NOB feature was also used on the D-Lite globe but never on the junior globe.

An entire book could be written on globes and chimneys. As with other antique glass items the lantern globe gives tell-tail clues to its age. Lantern globes do turn color, with purple as the most common. Bubbles in the glass indicate an early or possibly hand blown globe. Table 4.2, Dimensions of Common Globes, gives the height between the top and bottom shoulders (if any), top diameter, and bottom diameter to aid the collector in identification of lanterns and globes.

Plate 4.6
The classic No. 0 tubular on the right with a FITZALL LOC-NOB globe.

Plate 4.7
Note how the LOC-NOBs are used to hold the globe in the guard wires.

Plate 4.8
This is the classic movie lantern made from a 1936 Dietz Monarch and a home made tin reflector.

Table 4.2. Dimensions of Common Globes (sort by height)

Globe	Height	Top Diameter	Bottom Diameter
miniature	2.00 in. (5.0 cm)	2.00 in. (5.0 cm)	2.00 in. (5.0 cm)
Comet or No. 50	2.50 in. (6.4 cm)	2.38 in. (6.0 cm)	2.25 in. (6.4 cm)
The Original	2.63 in. (14.5 cm)	2.63 in. (14.5 cm)	2.38 in. (6.0 cm)
No. 40 fresnel	2.88 in. (14.6 cm)	3.00 in. (7.6 cm)	3.00 in. (7.6 cm)
kero	3.25 in. (8.3 cm)	3.25 in. (8.3 cm)	3.25 in. (8.3 cm)
little giant	3.50 in. (10.2 cm)	3.25 in. (8.3 cm)	2.75 in. (7.0 cm)
Vesta	4.00 in. (10.2 cm)	2.25 in. (5.7 cm)	2.75 in. (7.0 cm)
Air Pilot	4.25 in. (10.8 cm)	3.25 in. (8.3 cm)	3.38 in. (8.6 cm)
short (D-Lite)	4.25 in. (10.8 cm)	4.13 in. (10.5 cm)	3.13 in. (8.0 cm)
junior	5.13 in. (13.0 cm)	2.38 in. (6.0 cm)	2.75 in. (7.0 cm)
No. 39 or Vulcan	5.25 in. (13.5 cm)	2.50 in. (6.4 cm)	3.25 in. (8.3 cm)
Little Star U.S.	5.50 in. (14.0 cm)	2.44 in. (6.2 cm)	2.81 in. (7.1 cm)
Fitzall Loc-Nob	6.25 in. (15.9 cm)	2.75 in. (7.0 cm)	3.38 in. (8.6 cm)
No. 0 tubular	6.25 in. (15.9 cm)	2.75 in. (7.0 cm)	3.38 in. (8.6 cm)
No. 2	8.50 in. (21.6 cm)	4.25 in. (10.8 cm)	5.00 in. (12.7 cm)
No. 3	10.25 in. (26.0 cm)	5.75 in. (14.6 cm)	6.00 in. (15.2 cm)

Scene Spoilers

Knowledgeable collectors of any antique run the risk of having a pleasant period movie ruined by the jarring appearance of an inappropriate prop or set piece. This is more or less a problem depending on how observant the movie fan is to the non-story elements of costumes, sets, furniture, props, etc. The Art Director is responsible for the look and feel of the movie within the constraints of the budget.

Just a friendly warning. A brief review of the contents of this book will leave you with enough knowledge about lanterns to spot the inappropriate use of a modern lantern in a period movie. You will be able to tell your friends that the lanterns aren't correct for the period. But be warned, this will be more annoying to them, than informative.

Some of the more common lanterns (Monarch, Plate 8.3 and Blizzard, Plate 8.8) are seen in Western movies and TV programs again and again. The problem is, of course, these lanterns were designed and built about 50 years after the period. The classic American Western period begins around the time of the gold discovery in California in 1848 and continues until around the start of W.W.I in 1914. The lantern of the "Old West" was still a candle in a tin and glass cabinet built by the local "Smithy." Candles were the most abundant source of portable light in rural America from colonial times until the start of the 20th century.

The typical movie lantern is a 40s vintage Dietz Monarch with a cooking pan placed behind it as a reflector as shown in Plate 4.8. The effect of adding the reflector is to simulate the more costly Dietz Inspector of Plate 7.7.

The movie Western genre was popular in the 1930s, '40s, and '50s when the Monarch was still available at the corner hardware store. The Monarch is perhaps, the most abundant lantern today as well as one of the most recognizable. Its large fount and rounded top makes it standout from the rest. The abundance of Monarchs is due to their rugged construction, protective metallic blue finish, and that they were purchased in huge numbers by municipal street departments throughout the 1940s and 1950s. The Monarch is often seen with its original red globe and, on special order, the city's name embossed in the metal.

Since the Art Deco Monarchs were not build until 1936, they are totally inappropriate for period movies set before that date. Watch for Dietz Monarchs in TV reruns from "Gunsmoke" to "The Little House on the Prairie." Sam Drucker had a bunch of Monarchs hanging from the rafters of his General Store in "Petticoat Junction." "The Addams Family" had one in their dungeon. Norm has a Monarch above his workbench in "The New Yankee Workshop" on PBS. The 1936 Monarch is truly the kerosene star of Hollywood.

More recent TV Westerns, like "Doctor Quinn, Medicine Woman," rely on contemporary supplies. The good doctor's town seems to be filled with nothing but Chinese built Dietz Air Pilots.

Advertising

Now that we know the history and construction of kerosene lanterns it is easy to spot the good stuff from the run-of-the-mill. Plate 4.9 is an ad for an inexpensive Chinese reproduction. The ad says it is an 1840 hurricane lamp but, it is not an 1840 design and not a lamp. It is a knock-off of a Japanese design that has nothing to do with R. E. Dietz or any 1840 tooling. This lantern has bail ears and an inside lift. The only similarity between this lantern and anything from 1840 is this lantern also has an absorbent wick.

1840's Hurricane Lamp
Is Back In Three Colors

Named for exceptional performance under the most adverse weather conditions, these 10½" tall hurricane lamps are once again being manufactured using the original 1840 molds. Functional and decorative indoors and out, our hurricane lantern has a baked, colored enamel finish with shiny chrome-plated trim. It burns kerosene, common lamp oil, or citronella oil. Imported.
SPECIFY: Black (BLK), Fire Engine Red (RED), Hunter Green (DGR).
No.19535 Enamel Hurricane Lamp $17.95.

Plate 4.9
There is not much in this lantern that comes from 1840.

Chapter 5

The Lantern & Glass Companies

Lantern Manufacturers

The demand for oil and kerosene lanterns in the 19th century spawn untold numbers of lantern manufacturers. Practically every town had a tinsmith making lanterns for the local community. To get some idea of how wide spread this was just look how many Smiths there are in the phone book.

During the Civil War the rail transportation system was improved to the point that it was practical ship lanterns state-to-state. It was also during the war that the use of metal stamping machines to draw and press metal flourished in the U.S.

This was the age of the Industrial Revolution and lantern makers were there. Hundreds of small stamping companies appeared and just as the auto industry had its giant, Henry Ford, the lantern industry had Robert E. Dietz. The story of Robert Dietz's company is practically the history of kerosene lanterns in America.

Plate 5.1
The R. E. Dietz Company factory in New York city was built in 1887.

R. E. Dietz Company

Robert Edwin Dietz was born on January 5, 1818 in New York city. As a teenager Dietz worked in a hardware store and experimented with combinations of lamp fuels. In 1840 Dietz used his savings to buy a small oil lamp business in Brooklyn. The R. E. Dietz Company sold sperm oil, whale oil, camphene (distilled turpentine), glass lamps, candle sticks, and a few dead flame lanterns. Robert's brothers became partners in R. E. Dietz & Company in 1843. The business received a boost when P. T. Barnum paid Dietz to provide the lighting for one of his shows in 1850.

Coal oil (kerosene) was first distilled in quantity from coal in 1856 and Robert Dietz had a ready market for a cheap, bright burning fuel. Dietz was awarded a patent for a burner specially designed to burn the new oil. After Edwin Drake produced the first commercially successful oil well in 1859, the stage was set for an even cheaper source of kerosene.

During the 1860s, Civil War contracts, Robert's hard work, growth of railroads, and westward expansion made the lamp business a huge success. After the war ended, the cost of kerosene came down to a level where Dietz could sell lamps and lanterns to people who were still using candles.

In 1868, Robert Dietz met a salesman for Archer, Pancoast & Company named A. G. Smith. Archer, Pancoast owned the rights to produce and sell a new tubular lantern patented by John Irwin. A. G. Smith persuaded Robert Dietz to sell his interest in Dietz and Company and buy Archer, Pancoast. The new partnership, located in New York city, was called Dietz and Smith. Robert Dietz saw no future with his new partner and bought Smith out the next year.

Robert was on his own and changed the company name back to R. E. Dietz. The lantern business continued to be good and, in 1887, a new factory was built on the corner of Greenwich and Laught streets in New York. In 1894, Dietz retired and left his sons Frederick and John in charge. Robert E. Dietz passed away on September 19, 1897, at the age of 79.

Fire destroyed the ten-year-old factory in June 1897 and C. T. Ham offered to sell out to Dietz for $190,000. Instead, in February 1898, the board of directors secured controlling interest in

Plate 5.2
The R. E. Dietz factory in Syracuse became the center of activity when manufacturing ended at the N.Y. plant in 1931.

the Steam Gauge & Lantern Company of Syracuse, New York. The New York city factory was back in operation later that year. Additional factory buildings were built in Syracuse in 1905 and 1913. When Frederick passed away in 1915, John Dietz became President. Also in 1915 the equipment from the closed C. T. Ham Manufacturing Company was purchased.

Many lantern models were discontinued because of the Great Depression, and all manufacturing moved to Syracuse in 1931. However, the Dietz main office remained in NYC until 1952.

Embury Manufacturing closed December 31, 1952 and Dietz was ready to buy the equipment and even employee some of Embury's staff. To make more room for production of automotive products, Dietz moved much of the kerosene lantern production to a factory they purchased in Hong Kong in 1956. Much of the truck and safety light manufacturing, and some lantern production remained in Syracuse until 1962. Lantern sales were slow in the U.S. but seven models continued to sell world wide through the 60s, 70s and 80s. The Hong Kong property had become so valuable that when trade opened with China, the Hong Kong factory was sold and production was moved inside the border in 1986.

In 1992, a crippling United Auto Workers strike halted sales of truck lighting equipment and the Dietz family had lost interest in continuing the business. After five generations of Dietz family control, the name and equipment was sold to a Chinese industrialist.

New lanterns with the Dietz name continue to be stamped out in the Chinese factory and sold around the world, but a link to the great industrial revolution of the 19th century, was severed in 1992.

Plate 5.3
William Westlake, the inventive partner of John McGregor Adams.

Adams and Westlake (Adlake)

John McGregor Adams was born in Londonderry, New Hampshire on March 11, 1834. At the age of 19, Adams went to New York and got a job on the sales staff of Clark and Jesup. Adams was so successful he was sent to Chicago to run the sales office there. Adams formed a partnership with John Crerar and they joined Cross, Dane, and Westlake in 1874.

William Westlake was born in Cornwall, England on July 23, 1831. Westlake was a tinsmith for the LaCrosse & Milwaukee Railroad. In 1862 he invented the removable globe lantern, which he then manufactured in Chicago. The Great Chicago Fire of 1871 destroyed Westlake's factory but it was rebuilt and by 1873, the firm of Dane, Westlake and Covert had sales of $350,000.

On October 21, 1874, Mr. Adams' and Mr. Westlake's companies merged to create the Adams and Westlake Company (Adlake). Adams served as president, James F. Dane as vice president and Westlake was secretary. The resulting organization became the most successful railroad lantern company ever. Adlake merged with the Forsyth Brothers Company in 1899 to form the Curtain Supply Company which moved to Elkhart, Indiana in 1923. Adlake manufacturing moved from Chicago to Elkhart in 1927 where they continue to make railroad equipment today. In 1966, Adlake acquired another well known railroad lantern maker, the Lovell-Dressel Company.

Plate 5.4
The Adams & Westlake factory in Chicago has signs that read Railway Supplies and Brass Beds.

Defiance Lantern and Stamping Company

William Chamberlain Embury was born in Napanee, Ontario, Canada on December 17, 1873. As a young man Embury worked for a Canadian tin and lantern company named Kemp Manufacturing in Toronto. Embury moved to Rochester, New York and started the Defiance Lantern and Stamping Company in 1900. It is said that Embury chose the company name to declare his defiance of the big lantern firms like C. T. Ham and R. E. Dietz. Although he never beat Dietz in sales, Embury did live to see his next company, Embury Manufacturing, become one of the largest in the country.

Defiance Lantern and Stamping Company was financed by Embury's partners in Toronto and when they tried to force Embury to hire unskilled relatives in 1908, he left to start Embury Manufacturing. Embury probably did the right thing because Defiance never grew beyond a few dozen employees. During the teens and 20s Defiance made generic hot blast and cold blast barn lanterns plus a conventional No. 39 railroad lantern. Defiance was too small to survive the Stock Market crash of October 1929 and Embury was able to buy the dies and equipment in 1930.

Embury Manufacturing Company

At the age of 35, William C. Embury left the Defiance Lantern and Stamping Company to form the Embury Manufacturing Company on November 27, 1908, in Rochester, New York. Embury moved his factory to Warsaw, New York in 1911. His lantern company was so successful that Embury was able to purchase Defiance after the stock market crash of 1929.

Embury lanterns were often sold to wholesalers and municipalities without the Embury name, which makes identification more difficult.

William C. Embury retired in 1936 and the business continued under the direction of Phil, Fred, and William Jr. Use of the kerosene lantern was in decline after World War Two and Embury Manufacturing closed on the last day of 1952. The Dietz company purchased the equipment and moved it to the Dietz factory in Syracuse where some of the more popular lanterns continued to be made.

C. T. Ham Manufacturing Company

Charles Trafton Ham was born in North Berwick, Maine on September 25, 1824 and grew up to become a locomotive engineer on the Boston & Lowell Railroad. C. T. Ham had a successful career in railroading but he wanted to be his own boss. In 1871, at the age of 46, he became partners with J. H. Kelley. Their partnership eventually became the Buffalo Steam Gauge Company. When the steam gauge business merged with a lantern company in 1875, the name became Buffalo Steam Gauge and Lantern. The firm produced railroad headlights and markers in addition to steam gauges. The company and factory moved to Rochester in 1876. At the age of 61, C. T. Ham left S. G. & L. to start his own lantern company in Rochester. On May 1, 1886, C. T. Ham Mfg. Co. was founded to build tubular, street, square, headlights, railroad, commercial and vessel lamps and lanterns.

By 1894, C. T. Ham employed about 250 people and by 1912, 291. Charles passed away two days after his seventy-ninth birthday. The company closed in 1914 and was then purchased by the R. E. Dietz Company.

Plate 5.5
Image of Charles Trafton Ham taken about 1876.

Plate 5.6,
The factory of the C. T. Ham Manufacturing Company in Rochester, New York.

Handlan-Buck Manufacturing Co.

Alexander Hamilton Handlan was born in Wheeling, West Virginia on April 25, 1844. During the U.S. Civil War (1861-1865), Handlan learned bookkeeping in the Union Quartermaster Corp. In 1868, young Alexander secured a position with Myron M. Buck's company in St. Louis, Missouri. Alexander became a partner in M. M. Buck and Company in 1874, and bought Buck out in 1895. The company name was changed to Handlan-Buck Manufacturing Company in 1901. Alexander Handlan Jr. continued the firm that lasted until about 1960.

Steam Gauge and Lantern Company

The Buffalo Steam Gauge Company of Buffalo, New York, became the Buffalo Steam Gauge and Lantern Company in 1875 when it merged with the firm of Parmelee and Bonnell. In 1876 the factory moved to Rochester but the name was not changed until the company was incorporated in 1881. At that time a new factory was built on the Genesee River, the name was changed to Steam Gauge & Lantern Company, plus C. T. Ham and Colonel E. S. Jenney joined the corporation. The Colonel owned the rights to produce and sell Irwin's tubular lanterns in the western states. Charles T. Ham left S. G. & L. Co. to start his own lantern works in 1886.

Things were humming along as well as could be expected until the night of November 9, 1888. A fire started in the first floor shipping department of the seven story building. The fire quickly spread up the open stair wells trapping the night crew on the upper floors. Many threw themselves out the windows, smashing to the feet of horrified and helpless onlookers. Thirty eight perished immediately or days later from injuries received.

After the fire, the company moved to Syracuse, New York and resumed business. The R. E. Dietz Company bought a controlling interest in S. G. & L. Company after Dietz's New York city factory burned in 1896. The name of S. G. & L. Co. continued to be used on lanterns until it was phased out about 1900.

Wheeling Stamping Company

In 1878, Archibald Woods Paull Sr. started a metal stamping business called the Nail City Lantern Company in Wheeling, West Virginia. Their principle product was lanterns but they also produced numerous metal and glass items. Archibald Woods Paull II

Plate 5.7
The Steam Gauge and Lantern Works in Rochester, New York.

reorganized the company in 1897 and renamed it Wheeling Stamping Company. Wheeling made lanterns are marked "Paull's" and use plastic for the fount caps. Wheeling continued to make lanterns until 1946 when the lantern division was sold to the R. E. Dietz Company.

Perkins Marine Lamp Corporation

Frederick Perkins started Perkins Marine Lamp Corporation in 1916 to produce brass and hot galvanized iron lanterns and other ship fixtures. The Great Depression caused Perlins to reorganize and change the corporation name to Perkins Marine Lamp and Hardware. The company prospered through the 1930s, '40s, and '50s. The corporation moved to Miami, Florida, in 1960 and the name was changed to Perco, Incorporated.

Distributors, Jobbers, and Wholesalers

The names of many lantern distributors appear on lanterns made by the big manufacturers. Embury lanterns are often found with the name of a hardware or department store in place of the Embury name. It appears that anyone who placed an order big enough could specify their name on the lantern. Some of the names that show up are:

Belknap Hardware of Louisville, Kentucky. Shapleigh Hardware of St. Louis, Missouri. Ward's, Van Camp, Hercules, Gamble's, Coast-to-Coast Stores, Richards & Conover Hardware. Co. and Hibbard Spencer Bartlett, just to name a few.

Just about any city, county, state, Federal agency, utility, or private company could order lanterns with their identification stamped into the metal. Some are obvious, like:

U.S.M.C. (United States Marine Corps), P.G.&E. (Pacific Gas & Electric), D.W.&P. (Department of Water & Power), and U.S. Navy. Others are somewhat cryptic like Elgin or P.D.H.

Globe Makers

The investigation into lantern globes is more complex than the study of lanterns themselves because every lantern model has a variety of globe colors, details, and replacements. There are dozens of styles from countless lantern companies, glass works, and importers to stimulate the collector's palate. This book would not be complete without a discourse on the history and development of the classic globe types.

A Short History of Glass

Glass is made today using essentially the same process the Romans perfected over two thousand years ago. Glass is a fusing of silica using an alkali flux. The silica comes in the form of sand, quartz or flint and the alkali is soda or potash. Other ingredients that can be added include lime for stability, magnesium for clarity, lead oxide for brilliance, borax for hardness, and metal oxides for color. Glass is made by first finding a source of good quality silica from a river bed or by mining. Potash or soda (sodium carbonate) comes from burning certain species of plants, kelps, or wood. The silica is washed and heated to remove impurities and makes up 60 to 70% of the glass. About 20% of the glass is soda, which acts to fuse the silica and other ingredients. The lime, magnesium, and metal oxides make up the remaining 10 to 20%. If the ingredients are not mixed in the correct proportion the glass surface can deteriorate. When glass checks and flakes it is called crizzling. The ingredients are heated to about 1400° C and the impurities are skimmed off the surface. The glass is now ready to be worked, blown or pressed into molds. After the glass object is formed, annealing is done in a special oven called a lehr. The lehr allows the glass to cool very slowly so it does not turn brittle.

The English produced lanterns made with pale green glass lites around 1650, and by 1676, clear lead glass was being used. In The New World there were several unsuccessful attempts to start a glass industry as early as 1608. The first successful glass factory was founded in New Jersey by Casper Wistar in 1739. Wistar made window glass, bowls, jugs, vases, and cups. Later in the century, pattern molding into wood, stone, and clay was being employed to make decorative pieces. Glass production spread to Pennsylvania and Connecticut and by 1830 there were 90 glass factories in the U.S.

The New England Glass Company invented a process for machine molding in 1827. This advancement allowed for mass production of glass items to meet needs of a growing country. In the fixed globe era, before the Civil War, lantern globes were blown into wood or metal forms. Great skill was required to cut off the ends and grind the edges smooth. The molds were plain but the glass could be cut with decoration or the owner's name.

After the war, increased competition forced many smaller glass works to merge or fold. The high cost of glass molding machinery produced partnerships with the lantern makers. The lantern maker needed a reliable source of globes and the glass maker needed a dependable market. These partnerships allowed for capital investments in the heavy machinery that streamlined the manufacturing process. Two big winners in this dog-eat-dog lantern globe war are still in operation: Kopp and Corning.

Kopp Glass Company

Nicholas Kopp was the Chief Scientist for the Pittsburgh Lamp, Brass and Glass Company when they moved to Swissvale, Pennsylvania around 1900. Mr. Kopp was instrumental in the invention of the red selenium glass we are familiar with today. Before selenium glass, red was applied to the globe using a process called "flashing" where a thin coat of colored glass is applied to the inside of a clear globe as it is blown. The resulting color is not as consistent as Kopp's ruby glass.

In 1926, Nicholas Kopp took over the failed company's equipment and building in Swissvale. The Kopp Glass Company globes were sold direct to lantern manufacturers with the manufacturer's name cast in the glass. However the Kopp logo, a circle with a K inside, appears on some Kopp made globes. Kopp Glass, Inc. of Swissvale, PA believe they are the only U.S. company still making lantern globes.

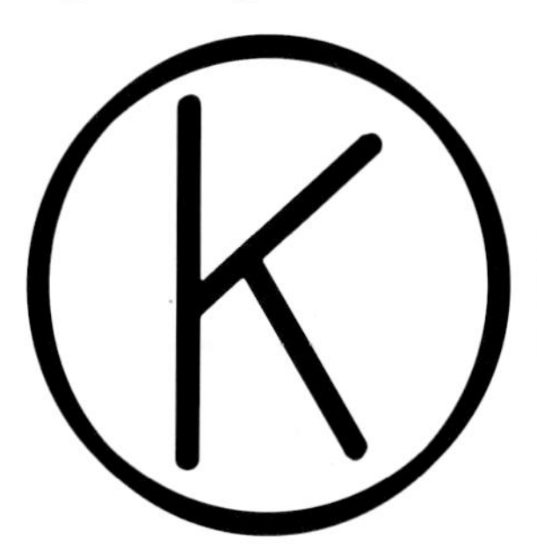

Plate 5.8
The Kopp Glass Company logo as seen on many lantern globes.

Corning Flint Glass Works

Due to the availability of wood for fuel, many glass factories were located around Pittsburg. In 1851, Amory Houghton, Sr. bought an interest in the Bay State Glass Company of Cambridge and three years later he founded the Union Glass Company in Somersville, Massachusetts. In 1864, Amory sold Union Glass and purchased the Brooklyn Flint Glass Manufactory. The company name was changed to Corning Flint Glassworks when, in 1868, Amory moved to be closer to the rail line in Corning, New York. The railroad now provided the coal for the glass furnace and Corning provided the railroads with signal lens and lantern globes.

Good as they were, lantern globes had a tendency to crack if unevenly heated by the flame or cooled by rain or snow. In 1908, Corning sponsored one of the first research laboratories in American industry and they went to work on the problem. In 1909, NoNex, the first low expansion glass, was produced for barn and railroad lanterns. NoNex led directly to the development of Pyrex® brand cookware in 1915. Corning heat resistant globes can be identified by their Trademark NX (for NoNex) inside a larger letter C.

Many other factories produced globes in North America and abroad, which makes for interesting research.

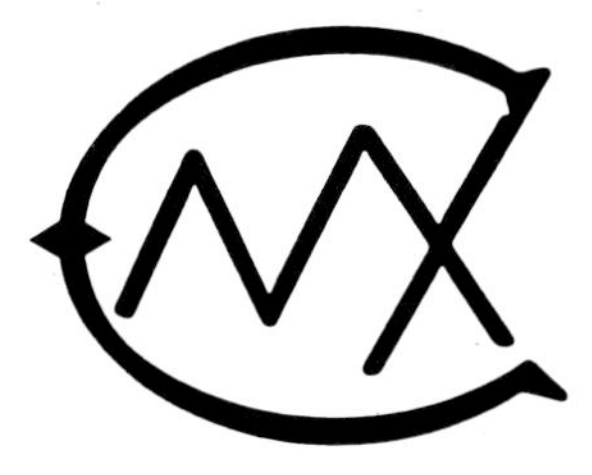

Plate 5.9
Trademark for NoNex glass globes made by Corning.

Chapter 6

All Solder Construction, 1830-1895

Coal Oil Comes of Age

This chapter presents the lantern examples that are constructed entirely by hand. In the earliest examples, tin smiths formed the parts from sheets of tin. Later, the parts were formed by large water or steam powered presses and then soldered by hand. Solder is a tin-lead alloy that melts at a temperature much lower than steel. Large copper irons are heated in a fire and used to solder metal parts of the lantern together by hand.

Chapters 6, 7, and 8 present lanterns with full descriptions grouped by approximate age from the oldest to the newest. Each lantern description includes:

- overall height excluding any top ring and handle,
- fount diameter (or width) to + 0.25 inch (+6 mm),
- all known markings,
- actual patent information marked on the lantern,
- approximate year(s) of manufacture,
- globe and wick details,
- and notes about related lanterns, manufacturer's history, innovations, and special features.

Table 6.1, Lanterns with Hand-Soldered Founts (sort by date)

PLATE	LANTERN MAKER	DATE	TYPE /STYLE	LANTERN MAKER	OVER-ALL HEIGHT	OVER-ALL WIDTH	GLOBETYPE	WICK (in.)
6.1	Kerosene Box Lantern	1830-1914	dead flame	various	various	various	glass panes	n/a
6.2	Fixed Globe Kerosene Lantern	1835-1890	dead flame	various	various	various	special	n/a
6.3	Dietz Pioneer Street Lantern	1880-1986	cold blast	R. E. Dietz Co.	25.0 in. (63.5 cm)	12.0 in. (30.5 cm)	No. 3	1.5
6.4	Universal World Standard	1880-1925	dead flame	Universal Metal	12.0 in. (30.5 cm)	6.75 in. (17.0 cm)	special	.625
6.5	S. G. & L. Co.L W	1885-1897	hot blast	S. G. & L. Co.	14.0 in. (35.6 cm)	5.75 in. (14.6 cm)	No. 0	.625
6.6	S. G. & L. Co.	1887-1897	hot blast	S. G. & . L. Co	13.5 in. (34.3 cm)	5.75 in. (14.6 cm)	No. 0	.625
6.7	SPCO Lamp	ca.1889	dead flame	Southern Pacific RR	12.75 in. (32.3 cm)	7.5 in. (19.0 cm)	10" lamp	.875
6.8	Winfield Mfg. Co.	ca.1891	hot blast	Winfield Mfg. Co.	13.75 in. (35.0 cm)	6.0 in. (15.2 cm)	No. 0	.625
6.9	S. G. & L. Co.Buckeye	1893-1898	hot blast	S. G. & L. Co.	13.75 in. (34.9 cm)	5.75 in. (14.6 cm)	No. 0	.625
6.10	C. T. Ham No. 2 (cold blast)	1893-1914	cold blast	C.T. Ham Co.	14.5 in. (36.8 cm)	7.0 in. (17.8 cm)	No. 0	.875
6.11	Ham's Clipper	1893-1914	hot blast	C. T. Ham Mfg Co.	13.75 in. (34.9 cm)	5.75 in. (14.6 cm)	No. 0	.625
6.12	C. T. Ham Gem	1893-1914	cold blast	C.T.Ham Co.	11.5 in. (29.2 cm)	5.5 in. (14.0 cm)	junior	.625
6.13	H.S.B. & Co. Bantie	ca.1895	cold blast	Hibbard Spencer	13.5 in. (34.3 cm)	5.75 in. (14.6 cm)	junior	.625
6.14	Dietz Victor	1897-1939	hot blast	R. E. Dietz Co.	13.5 in. (34.3 cm)	5.75 in. (14.6 cm)	No. 0	.625
6.15	Dietz Blizzard No.1	1898-1912	cold blast	R. E. Dietz Co.	14.0 in. (35.6 cm)	6.5 in. (16.5 cm)	No. 0	.625
6.16	C. T. Ham Clipper	1899-1914	hot blast	C. T. Ham Mfg Co.	13.25 in. (33.7 cm)	6.0 in. (15.2 cm)	No. 0	.625
6.17	Dietz Crystal	1899-1920	hot blast	R. E. Dietz Co.	14.0 in. (35.6 cm)	6.5 in. (16.5 cm)	No. 0	.625
6.18	Dietz Monarch (1900)	1900-1912	hot blast	R. E. Dietz Co.	13.25 in. (33.7 cm)	6.0 in. (15.2 cm)	No. 0	.625
6.19	Nail City Crank Tubular	1890-1897	hot blast	Nail City	14.0 in. (35.6 cm)	6.25 in. (15.9 cm)	No. 0	.625
6.20	Dietz O.K.	1892-1920	hot blast	R. E. Dietz Co.	13.5 in. (34.3 cm)	5.75 in. (14.6 cm)	No. 0	.625

Kerosene Box Lanterns

Plate 6.1 **Description:**

Oil lanterns came about shortly after the oil lamp. The first oil lantern was a candle box lantern with an oil pot in place of the candle. Box lanterns were easy to make. Any Smith could make a lantern from sheet steel and a standard window pane split in three equal pieces. When coal oil was produced in quantity in 1856, it soon became a less expensive substitute for sperm (whale) oil.

The first burner was just cord in a tube until Robert Dietz invented the flat wick in 1857. A popular burner of the day used two wicks set slightly apart to make a brighter flame. The wicks were adjusted before being lit until the sprocket wick adjuster was patented by Timothy J. Dietz in 1863.

The popularity of oil lanterns surged after 1865 when kerosene, refined from crude oil, became even cheaper than coal oil. Robert E. Dietz had been in business for 25 years and was selling a convertible model called the New Farm Lantern. The New Farm lantern had an insert for a candle holder, another for a sperm oil burner, and a third for coal oil.

If possible, the steel parts were japanned black to protect them from rust. Box lanterns could have three or four sides with glass in one or all of them. Wrist loops were popular until mid century but these gave way to the wire bail after the 1850s.

By 1890, R. E. Dietz was selling nine different Station Lanterns. Three have a square base and two are triangular lanterns that range in size from 15 to 21 inches tall. The remaining four are tubular models, three square boxes and one triangular box. All nine have a silvered reflector. When R. E. Dietz secured the rights to John Irwin's tubular patents, the tubes found their way into the simple box lantern.

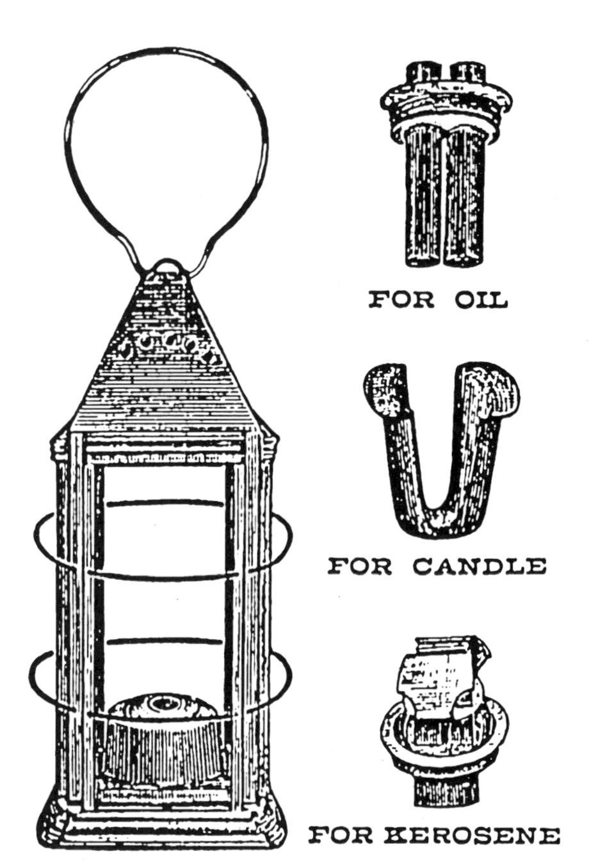

Plate 6.1a
The Dietz New Farm Lantern was one of the most popular due to its versatility.

Plate 6.1
Kerosene Box Lanterns, 1830-1914

JAPANNED STATION HOUSE LANTERNS—KEROSENE.

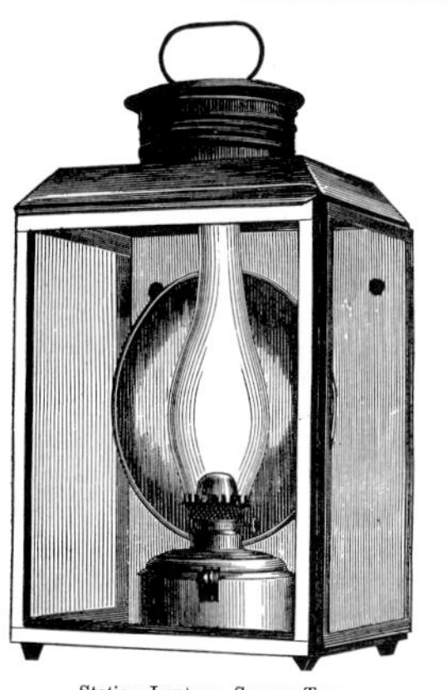

Station Lantern, Square Top.

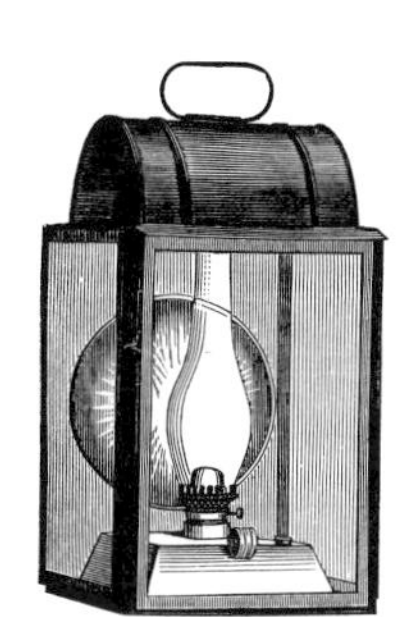

Station Lantern, Bow Top.

Boat Lantern.

Sexagon Sugar House Lantern.

Square Sugar House Lantern.

Street Lantern.

8×12 in. Square Flat Top, with Glass Reflector and Chimney, each $
10×14 in. Square Flat Top, with Glass Reflector and Chimney, "
12×16 in. Square Flat Top, with Glass Reflector and Chimney, "
9×13 in. Square Bow Top, with Glass Reflector and Chimney, "
9×13 in. Square Boat Lanterns, Looking Glass Wings...... "
Six-sided Sugar House Lanterns, Top Reflector............. "

Square Sugar House Lanterns, Top Reflector..............each, $
7×9 in. Square Japanned Street Lanterns ....per doz. $
5½ inch Square Japanned Street Lanterns, Plain .... "
Triangular Lanterns, Green Glass, for Oil................. "
Triangular Lanterns, Green Glass, for Kerosene............ "
☞ Signal, Ship, and all other kinds of Lanterns furnished to order.

PATENT CONVEX REFLECTOR LANTERN,
FOR BURNING COAL OIL OR KEROSENE WITHOUT A CHIMNEY.

It gives a pure white light, without the use of a chimney, and the flame can be regulated from the outside. It is neat and compact in form, and stands quick motion in any direction. By means of convex reflectors, the force of the flame is greatly increased, while they at the same time serve for the purpose of shades to the eyes.

Convex Reflector Lanterns......................per dozen, $

Burners for same..............................per dozen, $

Plate 6.1b
This early kerosene lantern is sometimes called the Excelsior.

Fixed Globe Kerosene Lanterns

Plate 6.2 **Description:**

The earliest kerosene lanterns with a globe are refered to as fixed globe lanterns because the crown and base are attached to the globe. They are dead flame lanterns with a pot removable from below for filling and wick adjustment. The globes can be described an egg, barrel, or onion shaped.

Remarks:

These early dead flame lanterns are typical of the fixed globe kerosene lantern in use during the second half of the 19th century. A similar whale oil lantern had been in use since the 1830s.

New technology is always expensive and the average person could not afford an oil lantern. Early oil lanterns were mostly purchased by companies like railroads and ship owners. There is no adjuster for the wick. The wick must be adjusted by hand before the lamp is lit.

The metal parts are usually tin coated steel and sometimes brass. The globe may be plain or marked for the railroad or steam ship company. Lanterns are constructed of tin coated sheet steel, cut formed and soldered by hand. Air holes are punched by hand and sometimes are in the shapes of stars, diamonds, and moons. The globes are mouth blown into steel molds just the way bottles were made. The bale is a hoop or wire attached to the top cover.

When the owner could afford it he had his name etched into the globe as often seen on early railroad lanterns.

Plate 6.2
Fixed Globe Kerosene Lanterns, 1835-1890. Each is hand made so no two are exactly alike.

LANTERNS, ETC.

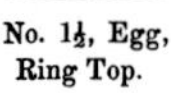

No. 1½, Egg, Ring Top.

No. 2, Flanged Egg, Ring Top.

No. 3, Egg, Bail Top.

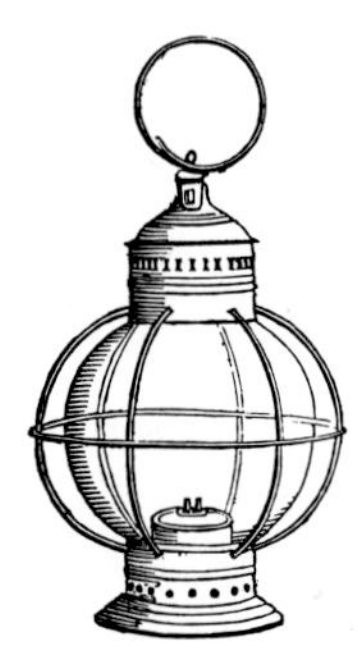

No. 1, Globe, Ring Top.

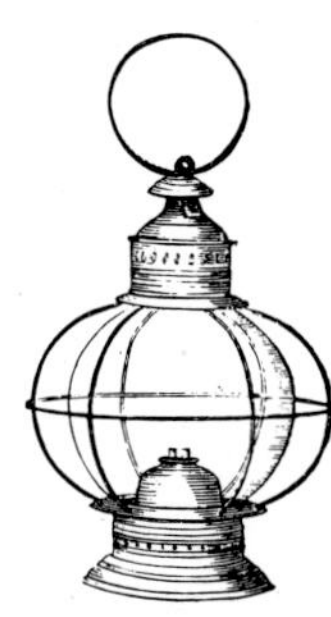

No. 2, Flanged Globe, Ring Top.

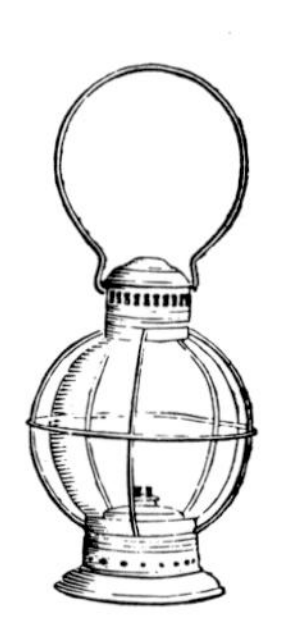

No. 3, Globe, Bail Top.

TIN LANTERNS—OIL.

Egg Pattern—Wire Bail Handle.

$					per dozen.
No.	1	1½	2	3	

Egg Pattern—Tin Ring Handle.

$					per dozen.
No.	1	1½	2	3	

Globe Pattern—Wire Bail Handle.

$						per dozen.
No.	1	2	2½	3	4	

Globe Pattern—Tin Ring Handle.

$				per dozen.
No.	1	2	3	

No. 1, Square Japanned....................per dozen, $

No. 5, Square Japanned....................per dozen, $

TIN LANTERNS—KEROSENE.

No. 2, Egg Pattern, Wire Bail Handle, Long Bottom........................per dozen, $

No. 2½, Globe Pattern, Wire Bail Handle, Long Bottom........................per dozen, $

No. 3, Globe Pattern, Wire Bail Handle, Long Bottom........................per dozen, $

No. 4, Globe Pattern, Wire Bail Handle, Long Bottom........................per dozen, $

No. 2, Egg Pattern, Wire Bail Handle, Ambrose........................per dozen, $

Clark's Patent, Egg, Wire Bail Handle.......per dozen, $

Breckenridge Patent, Egg, Wire Bail Handle..per dozen, $

Sargent's Patent, Egg, Wire Bail Handle.....per dozen, $

Plate 6.2a
Fixed globe lanterns with wire guards did not allow the globe to be removed. The fount and burner insert from the bottom.

WOODWARD'S PATENT

SELF-ADJUSTING COMBINATION LANTERNS—KEROSENE.

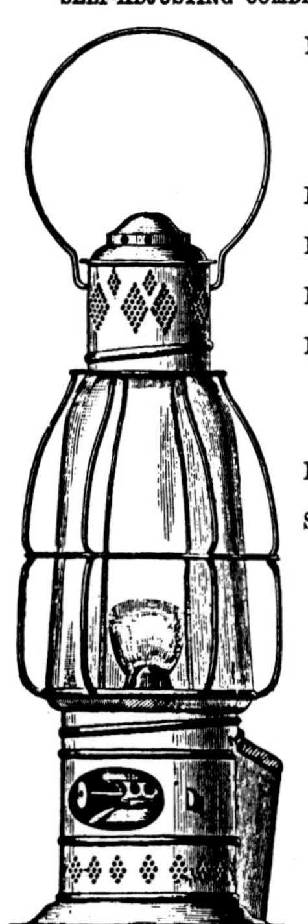

per dozen.

No. 1, with Match Box and Patent Slide and opening for lighting, regulating and extinguishing flame, without removing the Lamp, and Savage & Co's no Chimney Burner No. 2,.....$
No. 2, same as No. 1, without Wire Guard...............$
No. 5, same as No. 1, without Match Box................$
No. 6, same as No. 2, without Match Box...............$
No. 9, same as No. 1, shorter at the base, and without Match Box and opening in Band, with Savage & Co's No. 3 No-chimney Burner...............$
No. 10, same as No. 9, without Guard..................$
Savage & Company's No-chimney Burner (see Plate).........$

Savage & Co's
No Chimney Burner.

Plate 6.2b
This early lantern from 1865 has numerous options, including a match holder on the side.

Dietz No. 3 Tubular Street Lamp

Plate 6.3 **Description:**

The Dietz No. 3 Street Lamp is a tubular, cold blast street lantern, made of green painted steel. The large No. 3 globe has no guard and it is lifted by hand. A No. 3 globe is 10.75 inches (27.0 cm) tall and 8.5 inches (21.6 cm) wide. In 1906 the No. 3 became the popular Pioneer Street Lamp. The Pioneer uses a No. 3 burner with a 1.5 inch (3.8 cm) wick. Dietz made this lantern first in New York City, then in Syracuse, New York.

Plate 6.3
The Dietz Pioneer (1906-1944) is similar to the No. 3, (1880-1906), 25.0 in.H, 12.0 in.W, $220-$310 USD

Remarks:

The Dietz cold blast street lantern was called a Dietz No. 3 Globe Tubular before 1906 and Dietz Pioneer after 1906. It could be ordered in numerous different configurations. The burner could be ordered single or duplex. A duplex burner has two wicks next to one another. The finish could be brass, tin, or painted. Dietz claimed there were more of these lanterns in use than any other make.

This lantern could also be ordered in a hanging configuration without the post socket. The hanging version is called a No. 3 Globe Tubular Hanging Lantern. There was even a reflector available to make the No. 3 into a side light called a No. 3 Globe Side Tubular Lantern. Five different colors were available for the globe, clear, red, blue, green, or frosted.

An electrified version of the Dietz Pioneer was available in the mid 1980s and can be see all over the western streets of Disneyland and Walt Disney World.

Two similar lanterns called No. 2 Globe Tubular Hanging and No. 2 Globe Tubular Side Lanterns were made form 1888 to 1912. They used the No. 2 globe that is 9 inches (22.9 cm) tall and 7 inches (17.8 cm) wide.

Variations:

The finishes available included brass, tin, green and black. In addition to the kerosene burner, many Pioneers were sold wired for electricity.

Until about 1930, Adams and Westlake (Adlake) made a similar lantern called The Adlake No. 3 Globe Platform Lamp. C. T. Ham Manufacturing Company made a No. 9 Hanging Lamp, a No. 9 Street Lamp, and a smaller No. 4 Side Reflector.

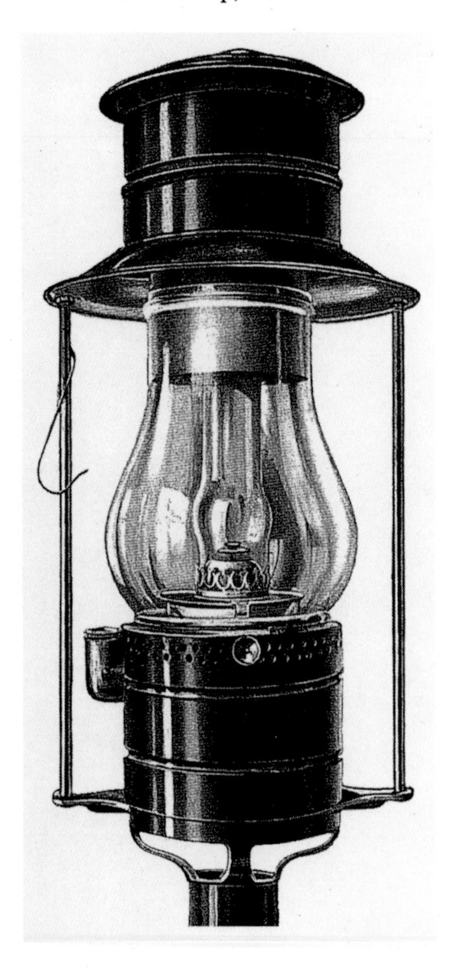

Plate 6.3a
Most companies made street lanterns. This dead flame Dressel was still available in 1926.

Universal World Standard Deck

Plate 6.4 **Description:**

The World Standard is an all brass, dead flame lantern with a closed base and double wire guard. The brass drop pot has a brass Vortex kerosene burner with a 0.625 inch (1.6 cm) wick. The globe is 3.5 inches (8.9 cm) diameter at the bottom, 5.25 inches (13.3 cm) tall, and 2.75 inches (7.0 cm) diameter at the top. The construction is all solder and the finish is polished brass. This lantern was made by Universal Metal Spinning and Stamping in New York City, U.S.A.

Markings: (stamped in crown)

THE UNIVERSAL METAL SPINNING
& STAMPING CO.
NEW YORK, N. Y.

Patent Information: (none)

Plate 6.4
Universal World Standard, 1880-1925, 12.0 in.H, 6.75 in.W, $220-$310 USD

Remarks:

This simple, dead flame lantern was made by at least three different companies about the same time, Perko, Dietz, and Universal. This happened because the design and specification was owned by the U.S. Navy. Suppliers would bid on government contracts to build the World Standard Deck lantern and one or more would get the contract. This lantern was used through World War One.

At the time lanterns like this were made, just about every lantern company had their own globe designs. By 1912 there were standard globes and most of the odd ones were no longer used. However this Universal lantern used an early globe rather than one of the designs that later became popular.

Most companies made just a few lanterns of brass for every hundred made of steel. Brass lanterns are highly sought after by collectors.

Universal Metal Spinning and Stamping Company first appears around 1910 and is gone by 1923. Metal spinning is a process usually reserved for soft metals like copper, brass, and aluminum. A flat sheet of metal is attached to a spinning die of the desired shape. As the stock spins, it is hand pressed with wooden tools to conform to the shape of the die. This method takes time and effort but parts can be made using a simple lathe and one piece die rather than a large press and two part dies.

Dietz made the same lantern with a ruby globe. The Dietz lantern is identical except for a Dietz logo at the top, a ruby globe, and two rings on the base. The tie down rings are a good indication of a lantern made for nautical use.

Plate 6.4a
The Dietz World Standard could be ordered with a sliding tin blackout cover on the globe.

Plate 6.4b
This lantern is made by Universal Metal Spinning of New York City.

Steam Gauge and Lantern Co. L W

Plate 6.5 **Description:**

Thc L W is a hot blast lantern that has a small brass fount cap, brass No. 1 burner, round tubes, top lift, and a small, flat fount. The correct globe is the early style No. 0. Finish is most likely tin or japanned black. This lantern is typical of the simple lanterns of the 1870s and 1880s.

Markings: (embossed on fount)

S. G. & L. Co.
L W

Patent Information: (on side tube)

PATENTED

May	28	'71
June	29	'80
July	28	'81
Mar	31	'85

Plate 6.5
S. G. & L. Co. No. 0, 1887-1897, 13.5 in.H, 5.75 in.W, $220-$310 USD

Remarks:

In 1868, John Irwin was awarded the patent for the tubular lantern. Most patents are for construction details like a fold in the metal for added strength or one tube that slips inside another. Patents seldom cause the broad sweeping changes that Irwin's tubular patent did.

In 1876 the Buffalo Steam Gauge & Lantern Company moved to Rochester. In 1881 they reorganized as S. G. & L. Co. and built a factory on the Genesee River to provide the power for the steel presses. One of the partners, Colonel E. S. Jenney, owned the right to sell tubular lanterns in the western states. R. E. Dietz Co. owned the rights for the Eastern and Central states.

One of the principle investors and President was Charles Trafton Ham, who later started a lantern company of his own in 1886.

The meaning of the letters L W on the fount are unclear.

Variations:

The S. G. & L. Company offered a complete line of railroad lighting products including a traction (trolley) headlight, locomotive and caboose marker lights, switch stand lanterns, cab lamps, and hand lanterns.

S. G. & L. also produced a square tube, dash lantern called Buckeye as shown in Plate 6.9. They also made several sizes of flat glass, kerosene, tubular, box lanterns. These lanterns had the usual three or four panes of glass but also tubes that returned hot exhaust to the burner.

Plate 6.5a
Kerosene headlight of the type made by S. G. & L. Co. and others.

Steam Gauge and Lantern Co. No. 0

Plate 6.6 **Description:**

This hot blast lantern has a small, one piece, brass fount cap and brass, No. 1 size burner that uses a 0.625 inch (1.6 cm) wick. The balance of the lantern is black, japanned steel. As with many early tubular lanterns, this one has square tubes, a top lift, and a small, flat fount. The correct glass is the early style No. 0, pear shaped globe protected by a slip on, two wire globe guard.

Markings: (on fount)

S. G. & L. Co. No. 0

Patent Information: (on center tube)

PATENTED
JULY 23 '71
JULY 26 '71
MARCH31 '85
MAY 24 '87

Plate 6.6
S. G. & L. Co. No. 0, 1887-1897, 13.5 in.H, 5.75 in.W, $220-$310 USD

Remarks:

The No. 0 tubular is a simple, inexpensive unit that must have been a good seller. This lantern is typical of the early hot blast lanterns built using John Irwin's tubular lantern patent. The side tubes carry hot air down to the burner where it fans the flame, creating a brighter light. The patents on this lantern reflect other innovations like the removable globe guard.

The name of this lantern is embossed on the fount. In this case No. 0 refers to the No. 0 globe that was relatively new at the time. In the 1880s, most oil lanterns were still the square box lantern with flat glass lights (Plate 6.1).

At 7:15 p.m. on November 9th, 1888, a fire started in the packing room of the 7 story, 7 year old building. The building had open stair wells and an open elevator shaft that helped the fire spread quickly. A special night shift was working to catch up after the election day shut down. The horse drawn fire equipment could not gain access to three sides of the burning building due to the river, a narrow street, and two other buildings. The employees were trapped on the upper floors, killing 38. The fire still ranks as one of the worst disasters in Rochester history.

With the help of other lantern makers, Steam Gauge was able to fill their orders and rebuild the business in Syracuse. The Dietz factory in New York city burned on June 23, 1897 so the Board of Directors purchased a controlling interest in S. G. & L. Co. The New York factory was rebuilt and both product lines were kept for a time (see the burner of the Dietz Victor in Plate 6.14). Eventually the Steam Gauge name was dropped.

Variations:

S. G. & L. Co. offered a dash lantern (see Plate 6.9, Buckeye Dash) that later became a mainstay of the R. E. Dietz line. S. G. & L. also had a removable tin reflector that slips over the tubes of the No. 0. The company also made several styles of boat and railroad lanterns.

Plate 6.6a

SPCO Station Lamp

Plate 6.7 **Description:**

The SPCO Station Lamp is a simple tin coated, steel fount and reflector with a standard lamp burner and chimney. This dead flame side lamp has no fount cap so the burner must be removed to fill. The height of the lamp is such that a 10 inch (25.4 cm) or taller chimney must be used. The reflector is embossed with the owners initials, in this case, Southern Pacific Company. A standard No. 2 lamp burner uses a 0.875 inch (2.2 cm) wick and a 3 inch (7.6 cm) diameter chimney.

Markings: (embossed on reflector)

SPCO

(alternate embossed on reflector)

UP (or other railroad)

(lamp burners are marked with the burner makers identification)

Patent Information: (none)

Plate 6.7
SPCO Lamp, ca., 1889, 12.75 in.H, 7.5 in.W, $50-$100 USD

Remarks:

This is a very common type of inexpensive lamp made in great quantities for the Southern Pacific, Union Pacific, and other railroads. It's made to sit on a small shelf in the depot.

This lamp is included in a lantern book because these tin lamps are often overlooked by the household lamp and railroad lantern collectors. It is made of tin, for rugged use by the railroads. The tin is formed to be a reflector, heat shield, and dust cover. Some versions have a soot guard, or crown, suspended over the top of the chimney. Parts are pressed on simple dies and assembled entirely by hand.

This simple lamp needed no patent protection as its construction was already in the public domain. The Southern Pacific Railroad had blueprints for standard stations, structures, and fittings. This lamp was undoubtedly made, as needed, locally by numerous tin smiths using Southern Pacific's standard plans.

Variations:

Minor variations are found from lamp to lamp, supporting the multiple supplier theory. The tops of some versions are cut, riveted, and soldered to form a curve. Some have a hole in the back for hanging on a nail. The UP version has a small, two part fill cap. Other lamps have a metal soot shield suspended over the chimney.

Plate 6.7a
Dressel No. 528 Bunk Car Lamp, 12 inches tall. ***From the collection of Scott E. Schifer***

Plate 6.7b
Catalogue cut of Dressel's Bunk Car Lamp.

Winfield Manufacturing Company Standard

Plate 6.8 **Description:**

The STANDARD hot blast lantern has a brass, No. 1 burner with a 0.625 inch (1.6 cm) wick and a brass, one piece fount cap. The balance of the lantern is tin coated steel. The globe lift is on the globe plate and the fount has a flat top. The glass in this lantern is a clear, No. 0 globe. Construction is all hand soldered. The bale is a loop attached to the center tube.

Markings: (embossed on the globe)

Winfield Mfg.Co.
Warren O.
STANDARD (See Remarks)

Patent Information: (stamped on center tube)

PAT. June 24, 1890

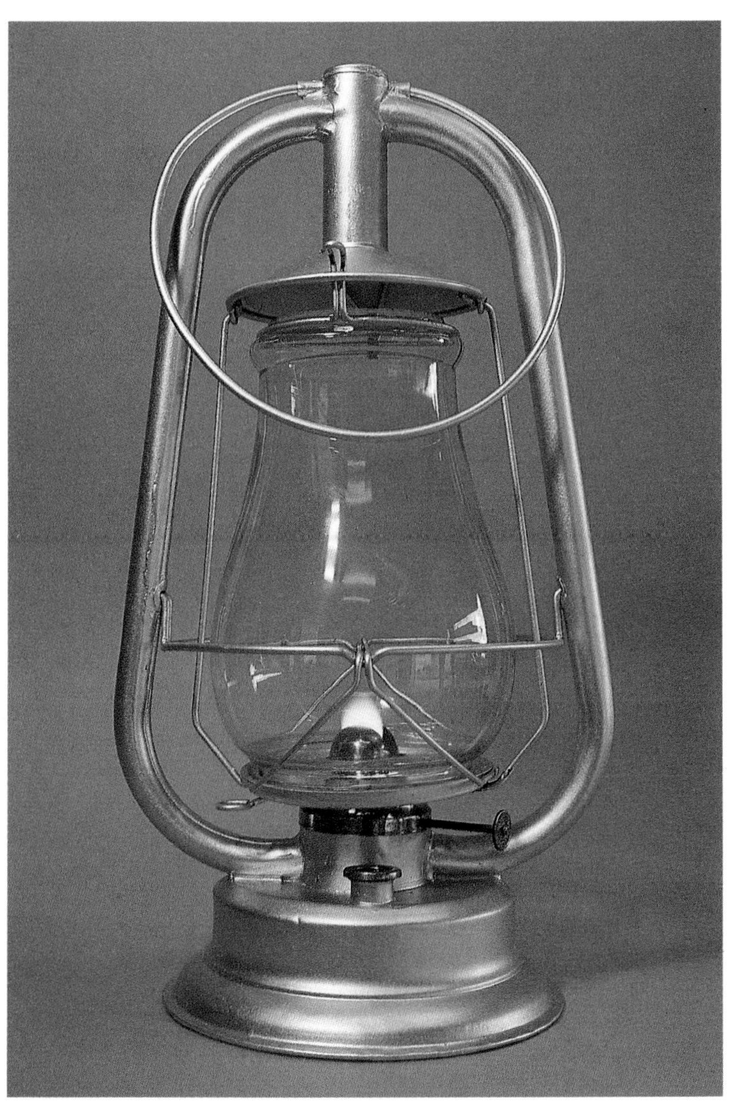

Plate 6.8
Winfield Mfg. Co. Standard, ca., 1891, 13.75 in.(35.0 cm) 6.0 in.(15.2 cm) $200-$290 USD

Remarks:

This early hot blast is unusual in that it has no embossed identification of manufacturer or country of origin. The only marking is the patent date stamped on the center tube. However, research and luck can fill in some of the blanks. While doing research I came across this Trade Card (Plate 6.8a) which shows this unusual lantern. The card is from a distributor located in Zanesville, Ohio. The lantern's name and maker are shown on the globe only. Winfield Manufacturing was located in Warren, Ohio from 1881 to the 1920s.

John Irwin's patents for the hot blast design ran out on May 4, 1886. Many small metal companies were free to manufacture hot blast lanterns without fear of harassment. However, the patent for the side lift was still in force. The lift on this lantern is unusual in the way the guard is hinged to the globe plate and folds when the globe is raised. Perhaps this is the patent cited on the lantern.

Crimp fount bottoms and stamped tubes were the state-of-the-art by 1890. The design of this lantern was obsolete and out of style when it was made around 1900.

The hoop bail is not uncommon for the earliest tubular lanterns as they took their styling from the fixed globe lanterns of Plate 6.2.

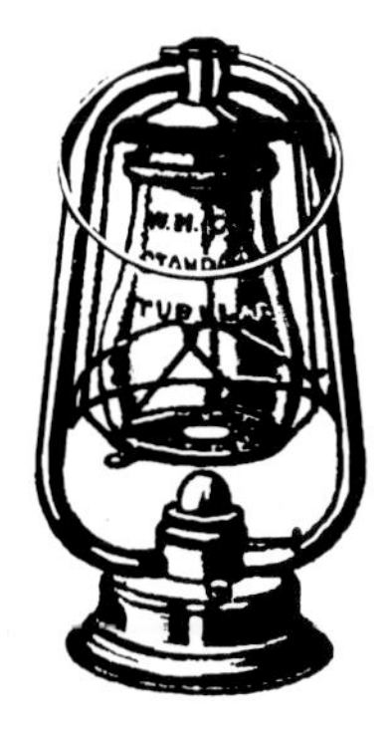

Plate 6.8a
Trade card showing the W. M. Co. Standard.

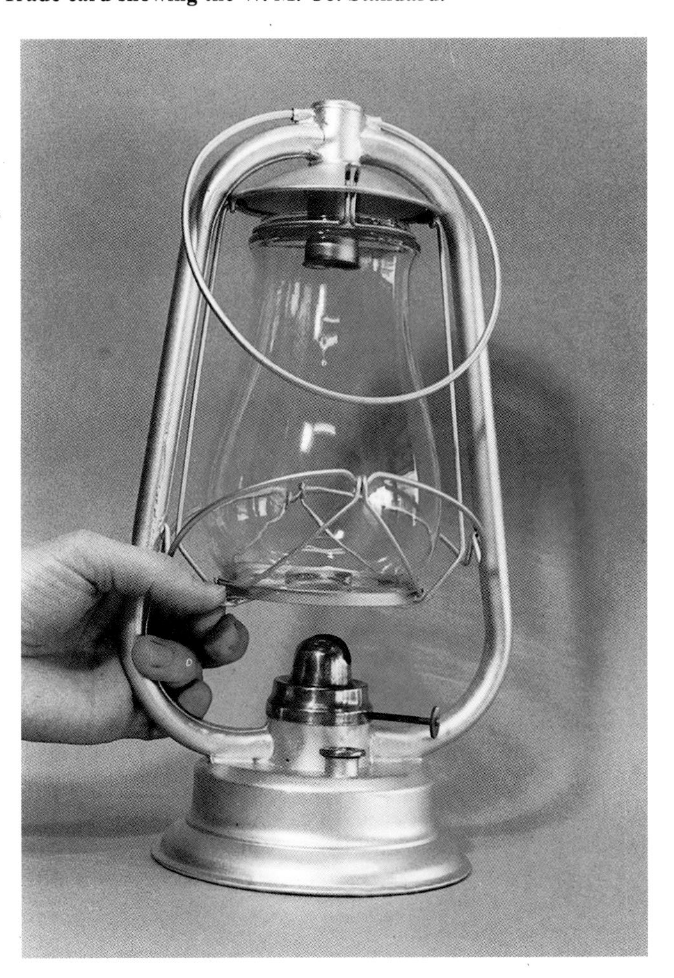

Plate 6.8b
The interesting arrangement of guard wires act as a hinge.

S. G. & L. Co. Buck Eye Dash

Plate 6.9 **Description:**

This Buck Eye Dash is steel and brass with many soldered parts. The reflector has a large spring loaded clip that can open to about 0.5 inch (1.27 cm) so the lantern can clip on the dash of a buggy. A bull's eye lens is attached to the globe plate, in front of the flame. The correct globe for this lantern is the early, pear shaped No. 0. Like most other dash lanterns this one was most likely finished in japanned black. The Buck Eye has a brass, No. 1 burner, 0.625 inch (1.6 cm) wick, and a small, brass fill cap.

Markings:(embossed on fount)

BUCK EYE
S. G. & L. Co

(embossed on globe)

S. G. & L. Co

Patent Information: (stamped on the center tube)

PATENTED
JUL - 28 - 81
MCH - 13 - 83
MCH - 31 - 85
MAY - 24 - 87
NOV - 15 - 87
??? - 13 - 88
??? - 24 - 88
??? - ?? - 88
??? - 21 - 88
??? - 8 - 90
??? - 21 - 90
??? - 24 - 91
MCH - 21 - 93
??? - ??? - 93

Remarks:

Colonel E. Jenney gathered $250,000 from investors to buy a lantern company named Dennis and Wheeler. The Dennis and Wheeler Company had the rights from John Irwin to sell hot blast lanterns west of the Mississippi. The Colonel also offered to buy the eastern U. S. rights from Robert E. Dietz, but Dietz did not sell.

The Steam Gauge & Lantern Company was making this buck eye dash lantern when Fred Dietz bought a controlling interest in 1897. In 1898, Dietz introduced a similar dash lantern shown in Plate 7.2.

The Buck Eye Dash Lantern, as its name implies, is a headlight that clips to the dashboard of a buckboard or buggy. The bull's eye is just a magnifying lens designed to focus the light on the road. A bull's eye globe has a lens molded onto the glass as the globe is made. Because the Buckeye Dash lantern has the bull's eye attached to the globe plate at the bottom of the globe, the more expensive bull's eye globe is not needed.

Note that a dash lantern does not have a red lens in the reflector as it would not show when clipped to a buggy dashboard. A wagon lantern (Plate 7.16) is similar to a dash lantern but mounts at the side of the wagon so the ruby "tail light" is visible from the rear.

This lantern must have a host of innovations judging by the long list of patent dates stamped on the center tube.

Plate 6.9
S. G. & L. Co., Buckeye, 1893-1898, 13.75 in.(34.9 cm) 5.75 in.(14.6 cm) $290-$400 USD

Plate 6.9a
S.G.&L.Co. logo.

Plate 6.9b
Another view of the S. G. & L. Co. Buckeye.

C. T. Ham No. 2 Cold Blast

Plate 6.10 **Description:**

The C. T. Ham No. 2 cold blast is one of the largest lanterns to use a No. 0 globe. The No. 2 has a brass burner with a plain wick adjust and a 0.875 inch (2.2 cm) wick. It has a small, one piece, brass fount cap, outside lift, and has all solder construction. This lantern is one of the earliest examples of the cold blast technology R. E. Dietz introduced in 1880.

Markings: (embossed on fount)
C. T. HAM MFG. CO. No 2
(a later version of the same lantern)
C. T. Ham MFG CO. ROCHESTER NY USA
No. 2 COLD BLAST
Patent Information: (on air chamber)
PATENTED
Feb. 28, 88
Oct. 17, 93
Oct. ?? 93
PATENT APPLIED
FOR C.T.HAM
MFG. CO.

Plate 6.10
C. T. Ham No. 2 (cold blast), 1893-1914, 14.5 in.(36.8 cm) 7.0 in.(17.8 cm) $50-$60 USD

Remarks:

The correct globe is a Ham's No. 0 but these are difficult to find. More Ham lanterns have survived than Ham globes. Care must be taken when lighting these large burners as setting the flame too high on a cold globe will crack it. An unmarked, old style, fluted globe would also be appropriate.

Here is what the 1902 Sears Roebuck Catalogue says about that No. 2 Cold Blast:

> Cold Blast No. 2 Round Tube, Bottom Lift, Tin Tubular Lantern; No. 2 burner, 1-inch wick, No. 0 globe. Globe removable without taking off the guard. This lantern being made on the same principle as a street lamp, with wind break, makes it a superior lantern in its burning qualities. Very desirable for use in places where there are strong drafts of wind. It is especially adapted for use in mills and other places where there is considerable dust, as the burner will not clog up. The tubes are made from one piece, without elbows or joints. Gives a fine light. We have noticed when one of these lanterns goes to a town we get more orders from the same locality. Weight about 2 1/4 lbs. Each...80c

The art of exaggeration is not new to advertising as No. 2 wicks are described as 1 inch but actually measure only 0.875 inch wide.

This lantern is from the "Wild West" days and would be correct for any movie set between 1880 to 1920.

Although there was no original paint left on this example, C.T. Ham usually japanned his lanterns blue. Tin was also identified in some of Ham's advertisements.

Variations:

For $1.00, the 1902 Sears and Roebuck Catalogue offers the Ham No. 2 lantern in a dash version with a reflector and dash spring clip like the Buckeye Dash lantern (Plate 6.9). Finish should be japanned blue, or tin plated.

Ham's Cold Blast Tubular Lantern.

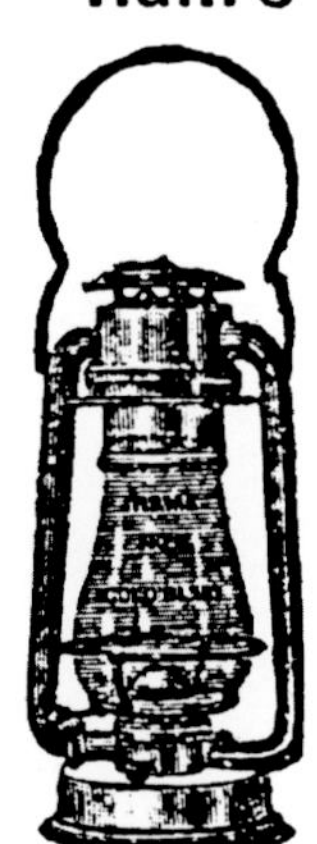

No. 23R7015 Cold Blast No. 2, Round Tube, Bottom Lift, Tin Tubular Lantern; No. 2 burner, 1-inch wick, No. 0 globe. Globe removable without taking off the guard. This lantern being made on the same principle as a street lamp, with **wind break,** makes it a superior lantern in its burning qualities. Very desirable for use in places where there are strong drafts of wind. It is especially **adapted for use in mills** and other places where there is considerable dust, **as the burner will not clog up.** The tubes are made from one piece, without elbow or joints. Gives a fine light. We have noticed when one of these lanterns goes to a town we get more orders from the same locality. Weight, 2¼ lbs. Each....80c

Plate 6.10a
Advertisement for Ham's Number 2 Cold Blast.

Ham's Clipper

Plate 6.11 **Description**:

Ham's Clipper is a hot blast, steel, tubular lantern with a small, one piece, brass fount cap, and a brass, No. 1 burner. The most striking feature is the odd combination of top lift and side lift. The top lift raises the globe only, while the side lift raises the globe and globe plate. The top lift was dropped as seen in the Clipper of Plate 6.16. No. 0 Clipper has a flat fount and blue japanned finish. This lantern must use a No. 0 globe or the top lift will not work. The Ham's Clipper was made by the C. T. Ham Mfg. Co. between 1893 and 1914 in Rochester, N.Y.

Markings: (embossed on fount)

No 0 HAM'S CLIPPER

(embossed on globe)

HAM

Patent Information: (on side tube)

PATENTED

AUG.	02	1880
SEP.	16	1890
MAR.	3	1891
OCT.	21	1893

Plate 6.11, Ham's Clipper, 1893-1914, 13.75 in.(34.9 cm) 5.75 in.(14.6 cm) $40-$50 USD

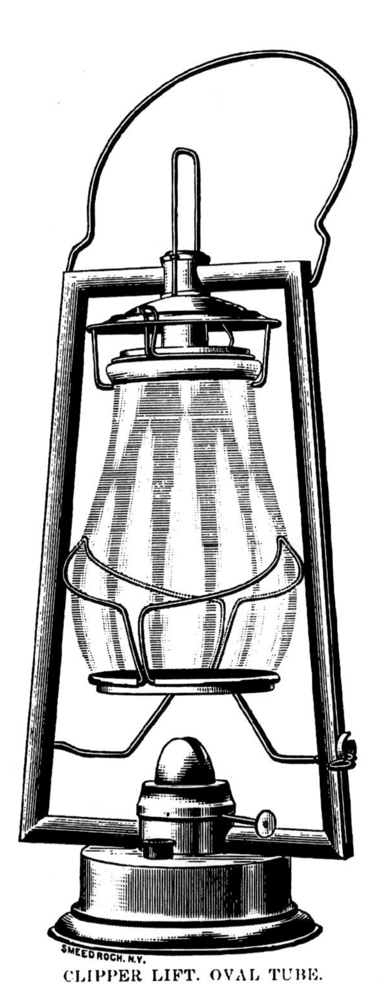

Plate 6.11a
The advertising called the tubes oval but they actually have a "D" cross section.

Remarks:

The No. 0 Ham's Clipper is an inexpensive tubular for home and farm use. This lantern's tubes have an unusual 'D' shaped cross section that was dropped in favor of round tubes by 1899.

Charles Trafton Ham died in 1903 and his son, George William Ham, continued the company until it was closed in 1914. The dies and equipment were bought by R. E. Dietz Company in 1915.

As with most lantcrn companies, Ham also manufactured dead flame railroad style lanterns, a No. 3 Globe Street Lamp, a Side Reflector, and the Ham's No. 40 Searchlight. Other Ham tubular lanterns are shown in Plates 6.10 and 6.12.

Variations:

The C. T. Ham Manufacturing Co. made a wide variety of box, street, side, police, fire, marine, auto, railroad, hot and cold blast lanterns. The less expensive hand lanterns seem to be the most common.

Plate 6.12
C. T. Ham Gem, 1893-1914, 11.5 in.(29.2 cm) 5.5 in.(14.0 cm) $180-$400 USD

Ham's Gem Cold Blast

Plate 6.12 **Description:**

The C. T. Ham Manufacturing Company built this all brass, cold blast, junior size lantern that uses a brass, No. 1 burner with a 0.625 inch (1.6 cm) wick. A junior globe is correct for the Ham's Gem. This lantern has a small, one piece, brass fount cap, outside lift, and polished brass finish. Patent dates are stamped on the hand soldered bottom fount plate.

Markings: (embossed on fount)

HAM'S GEM COLD BLAST
ROCHESTER, N.Y. U.S.A.

Patent Information: (stamped on fount bottom)

PATENTED
Feb. 28, 88
Oct. 17, 93
Oct. 31 93
PATENT APPLIED
FOR C. T. HAM
MFG. CO.

Remarks:

The Ham's Gem is truly a gem. A junior size lantern made of solid, polished brass is a real find.

Because of its small size and polish, this lantern is intended for use indoors. Advertisements for these types of lanterns would suggest its use by women and children. It lacks the large flame and rugged construction required of a good barn lantern.

For more information on buffing and polishing brass refer to chapter 10.

Remarks:

The Gem is also available in tin coated steel and there is an optional, removable reflector for dash use. Notice the globe holding guard on the Gem and the Ham advertisement in Plate 6.11a.

Plate 6.12a
The Gem is an example of a very early domed fount.

Hibbard Spencer Bartlett & Co. Bantie

Plate 6.13 **Description:**
The Bantie is a junior size, cold blast, wagon lantern because of its side mount bracket and rear, 2 inch (5.1 cm), red lens. It has a small, one piece, brass fount cap, and brass, No. 1 burner with a 0.625 inch (1.6 cm) wick. The Bantie also has an inside lift. Construction is all hand soldering.
Markings: (embossed on front)
H.S.B. & Co. OVB BANTIE
Patent Information: (none)

Plate 6.12b
When found, the tarnished Ham Gem didn't look quite so special.

Plate 6.13
Hibbard Spencer Bartlett & Co. Bantie, ca., 1895, 13.5 in.(34.3 cm) 5.75 in.(14.6 cm) $290-$400 USD

The Hibbard Spencer Bartlett & Co. of Chicago, Illinois, was a hardware distributor. H.S.B. & Co. is known to have had their name put on Embury lanterns in the teens and 1920s. This lantern's construction gives no clue as to who may have actually made it. H.S.B. & Co. used the "Our Very Best" (O.V.B.) slogan on numerous tools they sold.

Bantie refers to the small size of the lantern. It was never intended for export as it has no country of origin.

The Bantie is an odd little cold blast wagon lantern. Its cold blast design and all solder construction date it between 1880 and 1905.

The correct globe for this lantern is the standard junior size globe with a magnifying lens attached to the globe plate or a bull's eye globe.

As on most side lanterns, this example has its holder on its left side. It has no dash clip on the back but it does have a rear red lens. Compare this lantern to the Prisco Junior Wagon of Plate 7.16 and the Dietz Junior Wagon of 7.26.

Traces of black paint were found under the burner that may be original. Black was the preferred color for all wagon lanterns to match the color of surreys and buggies.

Plate 6.13a
This H.S.B.&Co. dash lantern was made about 1920. It is marked O.V.B. No. 2 (15.0 in.(9.0 in.) $100-$150 USD

Dietz Victor

Plate 6.14 **Description:**

The Dietz Victor is a better quality lantern with the latest improvements. The Victor is mostly steel with a small, brass fill cap, brass burner, and rear lift. The correct globe is the fluted, early style "DIETZ No. 0 TUBULAR." The off center fill is unusual for a Dietz lantern. It has all solder construction, square tubes, and a bright tin finish. The No. 1 burner uses a 0.625 inch (1.6 cm) wick. The Victor was made in the Dietz's New York City factory then in the Syracuse factory.

Markings: (embossed on front)
DIETZ No 0
VICTOR
(embossed on bottom)
DIETZ DIETZ

Plate 6.14
Dietz Victor, 1897-1939, 13.5 in.(34.3 cm) 5.75 in.(14.6 cm) $30-$100 USD

Standard Lamps & Lanterns
(on burner knob)
Dietz & S.G. & L. CO.
(on the globe)
DIETZ
No. 0 TUBULAR
NEW YORK U.S.A.
Patent Information: (stamped on center tube)
PATENTED
July-25-81
Mar-13-83
Mar-31-88
Mar-13-88
Oct-2-88
Jan-27-91
Mar-13-93
Remarks:

The 1902 Sears Roebuck catalogue describes the Dietz Victor this way:

> For Kerosene. This is the most popular lantern on the market today. The crank at the side raises and lowers the globe and locks the burner in place when down. A late improvement on this lantern consists of a bead on the guard wire over which the crank moves, thus perfectly locking the globe frame and burner down. No. 1 burner, 0.625 inch wick, No. 0 globe. Weight about 2 1/4 lbs. Each...47c

The patent dates are stamped into the center tube above the globe. This lantern has an interesting Dietz logo embossed on fount bottom. Dietz must have discovered that advertising on the bottom was not worth the effort, as it is not found on other models.

In 1902, the side lift lever was identified as a "late improvement." The burner's wick adjust knob is embossed with "DIETZ & S.G.&L. CO." This implies this burner was made shortly after the Dietz merger with S.G.&L. in 1897 and the dropping of the S.G.&L. name around 1899.

Variations:

Victors made from 1910 to 1939 have a dome top fount and large fill cap. Other variations include a sheet metal reflector or hood for dash or inspection use. The Dietz Iron Clad, made from 1889 to 1915, is very similar to the Victor.

Plate 6.14a
This unique Dietz logo is stamped in the bottom of the Victor.

Dietz Blizzard No. 1

Plate 6.15 **Description:**

The Dietz Blizzard is a tinned steel, cold blast lantern with a hand soldered fount typical of 19th century lanterns. It has a small, one piece, brass fount cap, a No. 0 globe, outside lift, brass No. 1, domed burner with a 0.625 inch (1.6 cm) wick. The air tubes are two part tubular and smooth. The fount has a flat top.

Markings: (embossed on fount)
DIETZ NEW YORK U.S.A.
No. 1 BLIZZARD
(embossed on globe)
BLIZZARD
Patent Information: (none)

Plate 6.15
Dietz Blizzard No.1, 1898-1912, 14.0 in.(35.6 cm) 6.5 in.(16.5 cm) $160-$250 USD

Remarks:

In production from 1898 to 1912, this was the first of the Dietz Blizzard line that continues to this day. The 1898 Blizzard evolved into the No. 2 Blizzard of Plate 7.15, then into the Art Deco model in Plate 8.8, and finally back to the Blizzard No. 80 of Plate 8.17.

When this lantern was made, Dietz had the rebuilt factory in New York City and the former Steam Gauge and Lantern Company factory in Syracuse, New York. This 1898 example is stamped from sheet tin with steel dies. The parts are then hand assembled and hand soldered. Dietz had been using their steam powered presses to draw and stamp parts since 1872. Before steam power, all lantern works had to be built next to a river because the presses and other machines were powered by huge water wheels. Steam power is just another example of the coming of the machine age.

This first Dietz Blizzard used a No. 1 burner so it didn't produce the light of the later No. 2 Blizzards.

Variations:

Available as an overhead lamp with a tin reflector to direct light down. Nearly identical to the 1912 to 1919 Dietz No. 2 Crescent.

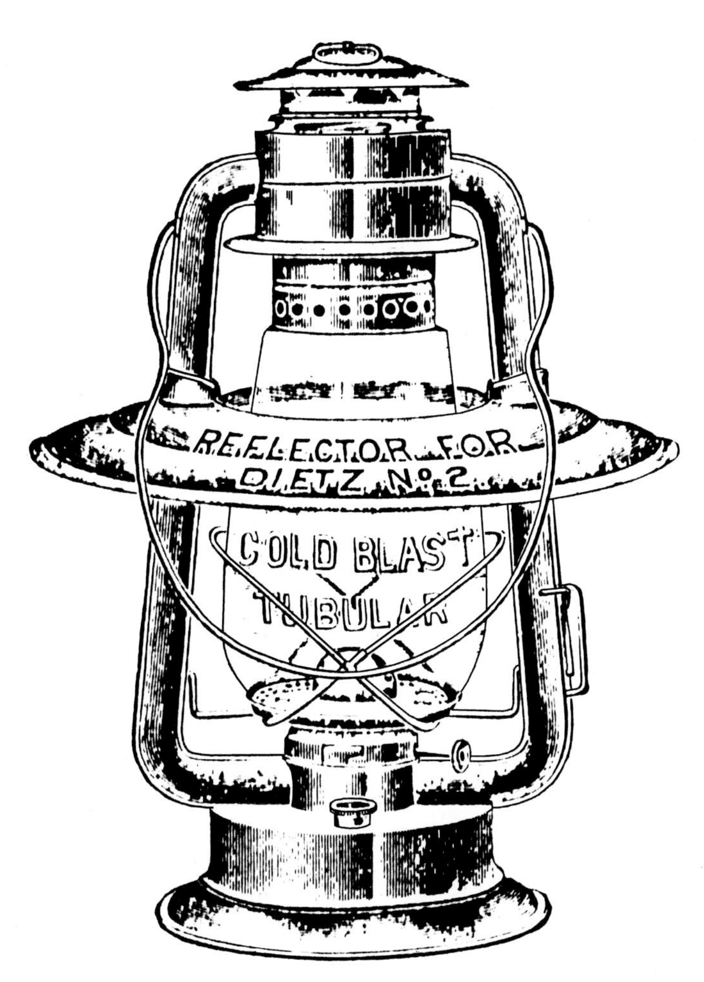

Plate 6.15a
The Blizzard shown with its optional reflector.

C. T. Ham Clipper No. 0

Plate 6.16 **Description:**

The Ham Clipper is your basic, tinned steel, hot blast lantern with a small, one piece, brass fount cap, brass No. 1 burner, outside lift, flat fount, and blue japanned finish. The correct globe is a pear shaped, "HAM," No. 0. The Clipper has all hand solder construction. The round air tubes were an improvement to the Clipper that continued until the end of the company in 1914.

Markings: (embossed on fount)

C. T. Ham MFG. CO. No 0 CLIPPER

Patent Information: (on side tube)

PATENTED

AUG.	02	1880
SEP.	16	1890
MAR.	3	1891
OCT.	31	1893
JULY	11	1899

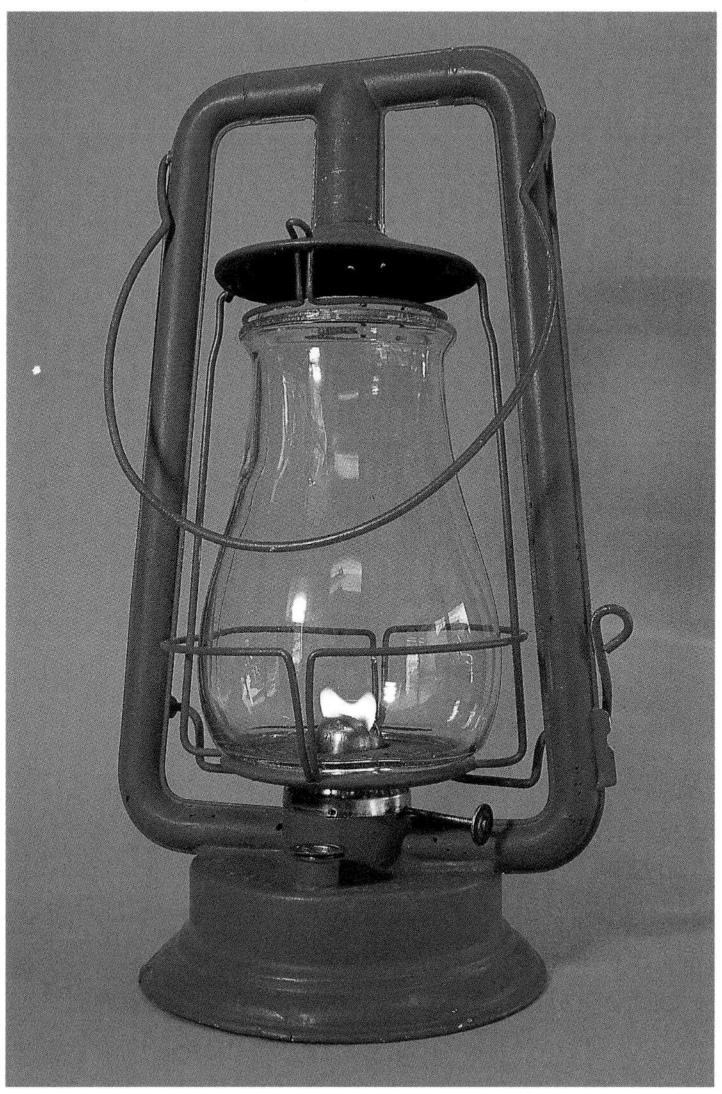

Plate 6.16
C. T. Ham Clipper, 1899-1914, 13.25 in.(33.7 cm) 6.0 in.(15.2 cm) $40-$50 USD

Remarks:

The Ham's No. 0 Clipper is simple, basic barn lantern of the type used by the thousands of dozens all over the world. (Lanterns were wholesaled in crates of one dozen.) Ham also sold burners and parts to other manufacturers. The Clipper was one of Ham's most successful models and is considered common.

This specimen was in poor condition when purchased. The rust and remaining paint was stripped revealing rust holes at the bottom of the hot air tubes. Lantern tubes rusted here because dirt, and dead bugs clog the tube hold moisture. If the holes are not repaired the lantern will not draft properly and smoke. If the holes are small, tin-lead solder would be an authentic repair and the value may not be affected. Plastic fillers should be avoided. On the other hand, if you don't plan to operate the lantern, the holes can be left as a badge of honor.

Charles Trafton Ham announced the formation of his new company on May 1, 1886, just three days before John Irwin's tubular lantern patents ran out. Charles had many years of experience as the President of the Buffalo Gauge & Lantern Company, and later with the Steam Gauge and Lantern Company.

The company was closed just before the start of World War One in 1914.

The older Ham Clipper with the "D" shape air tubes is shown in Plate 6.11. Other C. T. Ham Manufacturing Company lanterns are shown in Plates 6.10 and 6.12.

Variations:

The Clipper is available in "plain" (which is assumed to be tin), copper plate, nickel plate, and all brass.

Plate 6.16a
Ham logo of the period.

Dietz Crystal

Plate 6.17 **Description:**

The Dietz Crystal is a tin coated, steel, hot blast lantern with a glass fount. The glass fount is rust proof and replaceable. The top lift raises and lowers the globe and locks the burner in place when down. The fount fill cap is the small, brass type. The brass No. 1 burner is hinged, and domed, with a 0.625 inch (1.6 cm) wick. The Crystal uses a Dietz No. 0 tubular globe. The original finish was a better quality bright tin.

Markings: (embossed on fount)

DIETZ No 0 "CRYSTAL"

(cast in fount glass)

DIETZ CRYSTAL

(on burner knob)

DIETZ LANTERNS

(on the globe)

DIETZ No. 0
TUBULAR

Patent Information: (none)

Remarks:

The Dietz Crystal is a more expensive lantern because of the glass fount and bright tin finish. Bright tin is a thicker coating than gray tin.

The rust proof fount appealed to contractors who worked in dirt, mud, and snow. Rust is the enemy of any lantern that sits on

Plate 6.17
Dietz Crystal, 1899-1920, 14.0 in.(35.6 cm)
6.5 in.(16.5 cm) $200-$260 USD

the ground because moisture condenses under the base and causes the fount to rust through. Even when the rust hole is too small to be seen, the lantern will weep kerosene. The glass fount is not only rust proof but, should it be broken, it is replaceable.

The Crystal did not survive into the modern era as brass and copper founts did. They had the same rust protection without the hazards of breaking glass. The glass fount did have the advantage that the fuel level could be easily seen. The fount can be removed by pulling the fount guard wires from the slots in the air tubes.

The correct globe for all Crystals is the fluted, early style "DIETZ No. 0 TUBULAR."

Variations:

Around 1913, the lift was changed from the top lift to a rear lift. Compare to the Dietz Victor of Plate 6.14 and the Monarch of Plate 6.18.

Plate 6.17a
Detail of the removable glass fount.

Plate 6.17b
Early Dietz logo.

Dietz Monarch (1900)

Plate 6.18 **Description:**

The first Dietz Monarch is a rather plain hand lantern for home and farm use. It is an all steel, hot blast lantern with a small brass fount cap and a hinged, brass, domed, No. 1 burner using a 5.8 inch (1.6 cm) wick. The Monarch has the new rear lift, a flat top fount, and a hand soldered fount. The correct globe is the early style, pear shaped, "DIETZ No. 0 TUBULAR."

Markings: (embossed on fount)

Plate 6.18
Dietz Monarch (1900), 1900-1912, 13.25 in.(33.7 cm) 6.0 in.(15.2 cm) $30-$40 USD

DIETZ NEW YORK U.S.A.
MONARCH
(embossed on wick adjust)
DIETZ LANTERNS
(embossed on globe)
DIETZ
NO. 0 TUBULAR
NEW YORK U.S.A.

Patent Information: (on center tube)
PATENTED

MCH-	31	-??
AUG-	23	-85
MAY-	24	-87
APR-	21	-88
JUN-	??	-88
MCH-	??	-93
M -	??	- 00

Remarks:

At a time when most Dietz lanterns had square tubes, the Dietz Monarch was introduced with round tubes. The Monarch has a rear lift like the Victor in Plate 6.14, the Crystal in Plate 6.17, and the Dietz Royal. The rear lift on the Monarch is the only detail from these lanterns to survive to the present day.

The Monarch was built using two part tubes crimped together rather than soldered. This allowed lanterns to be built faster and the cost was reduced.

From the patent dates it is clear that this lantern was first made from 1900 until its replacement by the 1913 Monarch (Plate 7.19). Many other Dietz lanterns during this same time were made with crimp fount construction but the Monarch's fount was still being soldered by hand.

The correct finish for this lantern is japanned black or tin plated. The patent dates are stamped into the center tube above the globe. Question marks indicate unreadable information.

This lantern has the remains of human damage. A past owner punched holes around the perimeter of the base, presumably to nail the lantern down. These holes could be soldered and filed or left to tell their story.

Variations:

This lantern was the first member of what would become the most successful line of hot blast lanterns ever produced. Compare this lantern to the 1913 Monarch of Plate 7.19, the 1936 "Art Deco" version in Plate 8.3, and the re-release of the 1913 model in Plate 8.19.

Plate 6.18a
Detail of the Monarch fount. The holes indicate some former owner nailed the lantern down.

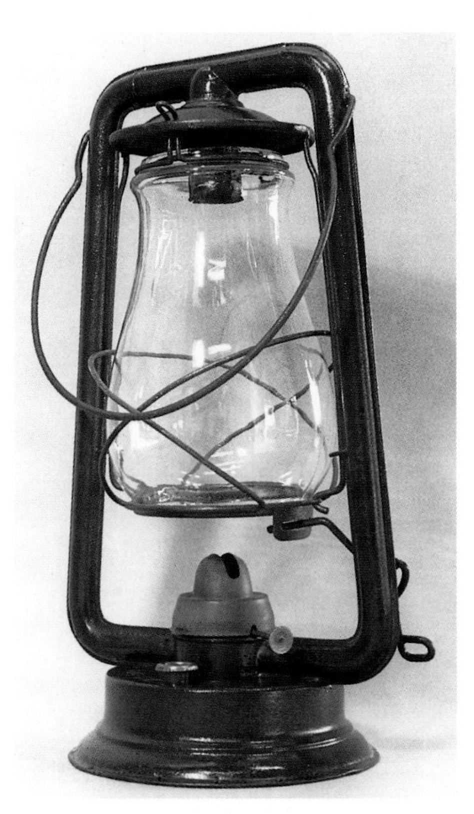

Plate 6.18b
Many things about the Monarch changed over the years but it always has a rear crank lift.

Nail City Crank Tubular

Plate 6.19 **Description:**

The Nail City Lantern Company stamped the Crank Tubular hot blast lantern from tin coated steel. The outside lift raises the globe up and back for access to the burner. The fount fill cap is a small, brass type. The No. 1 burner and dome are brass with a 0.625 inch (1.6 cm) wick. The Crank Tubular uses a N. C. L. Co. tubular, No. 0 globe. The original finish is bright tin.

Markings: (embossed on fount)
CRANK TUBULAR
N. C. L. Cos.

Patent Information: (none)

Remarks:

The air tubes of the Crank Tubular are stamped with corrugations and bent like a flexi straw. The unusual tube design did not prevent R. E. Dietz from filing a lawsuit against Nail City Lantern Company for infringement of John Irwin's tubular patent.

Archibald Woods Paull I, began the Nail City Lantern Company in 1877, and added the glass business in 1885. The company operated in Wheeling, West Virginia until it was taken over in 1889.

The age of this lantern can be well established by known changes at the Nail City factory. The ad in Plate 6.19a shows the same lantern but without the brass fill cap. The ad was published in an 1891 journal, and notes the fill is a new feature so the lantern of Plate 6.19 can not be much older than 1890. Archibald Woods Paull II changed the name of his father's company to Wheeling Stamping Company in 1897.

Wheeling Stamping continued to make lanterns like the LEADER (Plate 7.33) and the REGAL (Plate 7.34) until the lantern division was bought by R. E. Dietz Company in 1946.

Variations:

It is known that Nail City made a double globe lantern and some railroad style lanterns. Nail City's lantern production was estimated to be 180,000 in 1895, which may sound like a lot, but very few examples have survived.

Plate 6.19
Nail City Crank Tubular, 1890-1897, 14.0 in.(35.6 cm) 6.25 in.(15.9 cm) $200-$290 USD

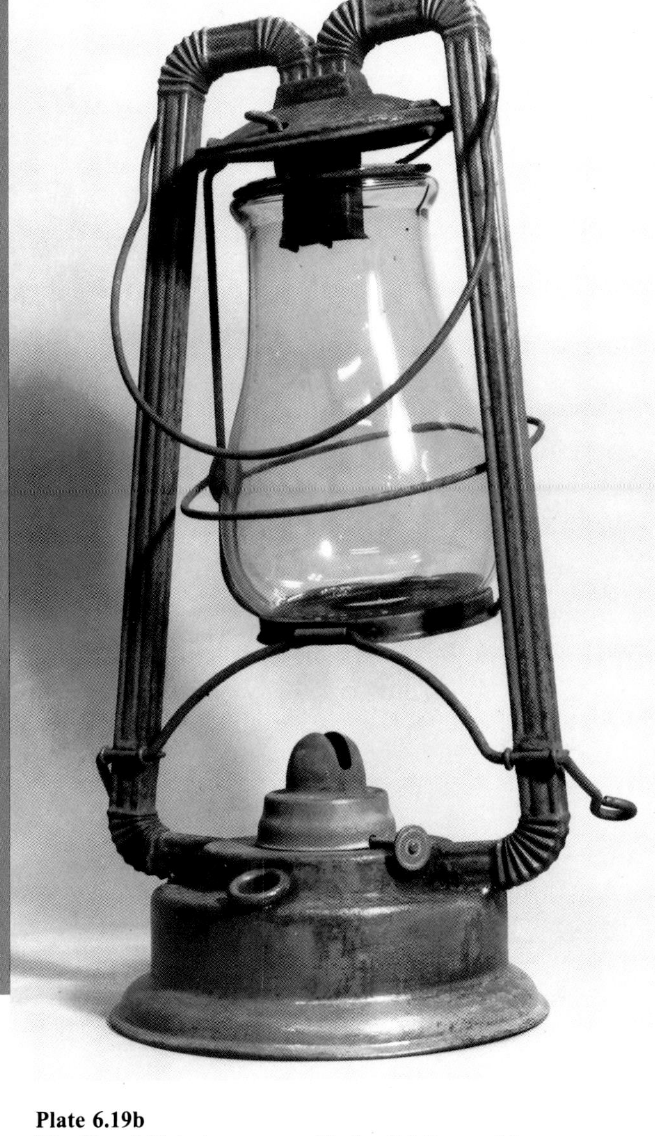

Plate 6.19b
The Crank Tubular opens wide for lighting and burner access.

Nail City Lantern Co.,

Manufacturers of the Celebrated "N. C. L. Co.'s"

Crank Tubular.

PERFECT IN EVERY RESPECT.

Mason Fruit Jars and Mason Porcelain Lined Trimmings. Trimmings furnished separate from Jars. All goods warranted strictly first class.

WHEELING, W VA

All our Crank Tubular Lanterns now have outside Brass Fillers.

Plate 6.19a
An 1891 advertisement for Nail City jars and lanterns.

Dietz O. K.

Plate 6.20 **Description:**

The O.K. hot blast, tubular lantern is mostly tin coated steel with a small brass fount cap and a brass, domed, No. 1 burner that uses a 0.625 inch (1.6 cm) wick. The O. K. has no lift but the burner plate is hinged to fold back. A front catch holds the globe, plate and burner in place during operation. This lantern has a flat top fount and all solder construction. The globe is an early style, pear shaped, "DIETZ No. 0 TUBULAR."

Markings: (embossed on fount)

DIETZ No 0
O K TUBULAR

(embossed on wick adjust)

DIETZ LANTERNS

(embossed on globe)

DIETZ
NO. 0 TUBULAR
NEW YORK U.S.A

Patent Information: (on center tube)

PATENTED
NOV 6 8?
MARCH13 88
88 7 ?
APRIL 22 90
OTHER PATENTS PENDING

Remarks:

After the recovery from the economic depression of 1893, Dietz offered a series of seven square tube, No. 0 globe, hot blast barn lanterns with different details and prices. The least expensive was the 58¢ Regular, which has no lift. The top lift, side lift and O.K. lanterns sold for 63¢. The Anti-Friction lantern has a screw lift and sold for 75¢ in 1899. Most expensive was the Crystal (Plate 6.17) and the Royal with its No. 2 burner and larger fount.

The side lift used on most other lanterns must have been preferred by customers as the O.K. was discontinued about 1920. The inside and rear lift eventually became standard on all lanterns.

The correct finish for this lantern is bright tin. The patent dates are stamped into the center tube above the globe. Question marks indicate unreadable information.

Variations:

After 1897, the Dietz O.K. has a double crimp fount, and after 1912 it sports a large fill. The same hinged globe plate is found on the Dietz Simplex hot blast lantern, which has a No. 2 burner and round tubes.

Plate 6.20
Dietz O.K., 1892-1920, 13.5 in.(34.3 cm) 5.75 in.(14.6 cm)
$160-$250 USD

Plate 6.20a
The O.K. globe folds back on its hinged burner plate.

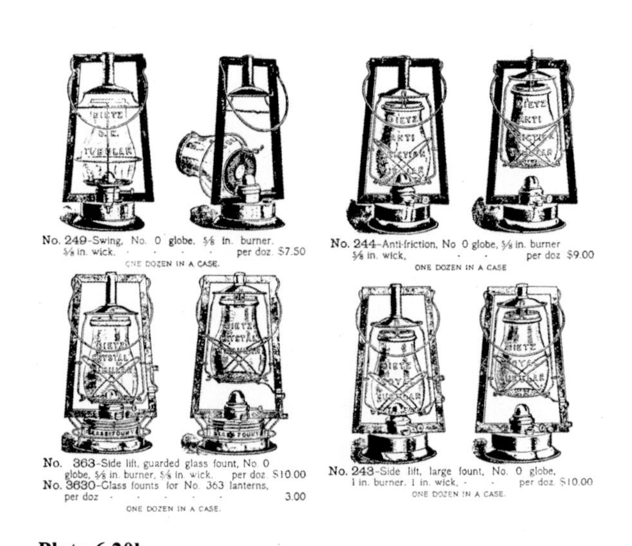

Plate 6.20b
A cut from the 1899 catalogue shows Dietz gave customers a number of choices in hot blast barn lanterns.

Chapter 7

Machine and Solder Construction

In this chapter, the lantern examples demonstrate where machine and hand assembly go hand-in-hand. By 1900, the dozen or so big players had secured the machinery to roll the fount seam rather than solder it by hand. In machine sealed founts, the bottom is attached and sealed by rolling the edges of the tin together. This construction method follows exactly, the improvements in tin food cans. When storage of food in tin cans began, they too were solder sealed. The use of tin-lead solder in food containers is now known to be a cause of lead poisoning in thousands of people. In the 1880s, the rolled edge we see on tin cans was developed. The same sealing machinery found its way into the lantern factories and is still the preferred method. The new methods meant all the air tubes were now stamped from a sheet of tin coated steel, then crimped in a press.

Yet other parts were still assembled by hand with solder. Solder was still being used to attach the tube assembly to the fount, but as designs were further refined, less and less hand solder processes were needed.

Table 7.1, Lanterns with Machine Crimp Founts & Solder (sort by date)

PLATE	LANTERN MAKER / NAME	DATE	TYPE/STYLE	LANTERN MAKER	OVER-ALL HEIGHT	OVER-ALL WIDTH	GLOBE TYPE	WICK(in.)
7.1	Dietz King Fire Department	1897-1940	hot blast	R. E. Dietz Co.	14.5 in.(36.8 cm)	5.75 in.(14.6 cm)	No. 0	.625
7.2	Dietz Buckeye Dash	1898-1946	hot blast reflector	R. E. Dietz Co.	13.25 in.(33.7 cm)	6.0 in.(15.2 cm)	No. 0	.625
7.3	Ham No. 2 (hot blast version)	1898-1914	hot blast	C. T. Ham Mfg Co.	14.0 in.(35.6 cm)	6.5 in.(16.5 cm)	No. 0	.875
7.4	Adlake Switch Stand	1899-1940	dead flame	Adams & Westlake	14.75 in.(37.5 cm)	7.0 in.(17.8 cm)	lens	.875
7.5	Dietz Hy-Lo	1899-1945	hot blast	R. E. Dietz Co.	13.5 in.(34.3 cm)	6.0 in.(15.2 cm)	No. 0	.625
7.6	Buggy/Auto Side Lamp	1900-1916	cold blast	J. W. Brown	11.75 in.(29.8 cm)	5.5 in.(14.0 cm)	glass	.625
7.7	Dietz ACME Inspector	1900-1954	hot blast reflector	R. E. Dietz Co.	14.75 in.(37.5 cm)	6.5 in.(16.5 cm)	LOC-NOB	.625
7.8	Hurwood Aladdin	1907-1915	cold blast	Hurwood Mfg. Co.	15.0 in.(38.1 cm)	6.75 in.(17.0 cm)	No. 0	.875
7.9	Dietz No. 39 Standard	1900-1944	dead flame	R. E. Dietz Co.	9.75 in.(24.8 cm)	6.0 in.(15.2 cm)	No. 39	.625
7.10	Handlan-Buck The Handlan	1901-1918	dead flame	Handlan-Buck	10.5 in.(26.7 cm)	7.0 in.(17.8 cm)	special	.625
7.11	Dietz Vesta	1906-1950	cold blast	R. E. Dietz Co.	10.0 in.(25.4 cm)	6.5 in.(16.5 cm)	vesta	.375
7.12	Dietz Beacon	1906-1945	cold blast	R. E. Dietz Co.	15.0 in.(38.1 cm)	6.0 in.(15.2 cm)	LOC-NOB	.875
7.13	Dietz Champion	1907-1930	cold blast	R. E. Dietz Co.	11.5 in.(29.2 cm)	7.75 in.(19.7 cm)	lens	.5
7.14	Embury Camlox	ca. 1911	cold blast	Embury Mfg. Co.	15.0 in.(38.1 cm)	6.75 in.(17.0 cm	No. 0	.875
7.15	Dietz Blizzard (1912)	1912-1945	cold blast	R. E. Dietz Co.	14.5 in.36.8 cm	6.75 in.17.0 cm	LOC-NOB	.875
7.16	Prisco Junior Wagon	1912-1920	cold blast	Pritchard Stamping	12.75 in.32.3 cm	5.5 in.14.0 cm)	junior	.625
7.17	Prisco No 477	ca.1912	cold blast	Pritchard Stamping	15.25 in.(38.7 cm)	6.75 in.(17.0 cm)	No. 0	.875
7.18	Dietz D-Lite (1913)	1913-1944	cold blast	R. E. Dietz Co.	13.25 in.(33.7 cm)	7.75 in.(19.7 cm)	short	.875
7.19	Dietz Monarch (1913)	1913-1950	hot blast	R. E. Dietz Co.	13.5 in.(34.3 cm)	6.0 in.(15.2 cm)	LOC-NOB	.625
7.20	Adlake Reliable Brakemen's	ca.1913	dead flame	Adams & Westlake	10.0 in.(25.4 cm)	7.0 in.(17.8 cm)	No. 39	.625
7.21	Adlake RR Semaphore	ca.1913	dead flame	Adams & Westlake	15.5 in.(39.0 cm)	5.25 in.(13.0 cm)	clear lens	.25
7.22	Warren STA-LIT	ca.1914	cold blast	Warren Stamping	14.0 in.(35.6 cm)	7.75 in.(19.7 cm)	short	.875
7.23	Buhl Little Giant	1914-1925	cold blast	Buhl Mfg.	13.0 in.(33.0 cm)	6.75 in.(17.0 cm)	short	.875
7.24	Defiance Perfect	1914-1930	hot blast	Defiance Lantern	13.25 in.(33.7 cm)	6.25 in.(15.9 cm)	No. 0	.625
7.25	Dietz Junior	1914-1956	cold blast	R. E. Dietz Co.	12.25 in.(31.0 cm)	5.25 in.(13.3 cm)	junior	.625
7.26	Dietz Junior Wagon	1914-1945	cold blast reflector	R. E. Dietz Co.	12.0 in.(30.5 cm)	5.25 in.(13.3 cm)	junior	.625

Table 7.1, Lanterns with Machine Crimp Founts & Solder (sort by date) (cont.)

PLATE	LANTERN MAKER / NAME	DATE	TYPE/STYLE	LANTERN MAKER	OVER-ALL HEIGHT	OVER-ALL WIDTH	GLOBETYPE	WICK(in.)
7.27	Dietz Eureka	1914-1930	dead flame	R. E. Dietz Co.	7.5 in.19.0 cm	5.0 in.12.7 cm	3 lens	.625
7.28	Embury No. 160 Supreme	1916-1952	cold blast	Embury Mfg. Co.	14.5 in.36.8 cm	7.25 in.18.4 cm	short	.875
7.29	Defiance No. 200	ca.1919	cold blast	Defiance L & S Co.	15.75 in.40.0 cm	7.5 in.19.0 cm	No. 0	.875
7.30	Embury No. 210 Supreme	1919-1938	hot blast	Embury Mfg. Co.	13.25 in.33.7 cm	6.0 in.15.2 cm	No. 0	.625
7.31	Embury Little Supreme	1920-1952	hot blast	Embury Mfg. Co.	12.75 in.32.3 cm	7.25"18.4 cm	wizard	.625
7.32	Dietz Little Giant	1920-1962	cold blast	R. E. Dietz Co.	11.75 in.29.8 cm	7.75 in.19.7 cm	wizard	.625
7.33	Wheeling Leader	1920-1946	cold blast	Wheeling Stamping	15.0 in.38.1 cm	7.75 in.19.7 cm	No. 0	.875
7.34	Wheeling Regal	1920-1946	hot blast	Wheeling Stamping	13.5 in.34.3 cm	6.75 in.17.0 cm	No. 0	.625
7.35	Perkins Perko	1920-1930	dead flame	Perkins Marine	10.5 in.26.7 cm	5.25 in.13.3 cm	special	.625

Dietz King Fire Department

Plate 7.1 **Description:**

The Dietz King is a nickel plated, brass lantern made for fire department use. The globe guard and draft shield is hinged at the sides and can be easily flipped up to light and extinguish. The correct globe is an early, fluted, DIETZ No 0 TUBULAR. The unusual, heavy bale and hanger is normal for fire department lanterns. The King has a one piece, brass fount cap, No. 1 burner, and a square (top) lift. These lanterns have a double wall fount and are noticeably heavier than others of similar size.

Markings: (embossed on fount)

DIETZ KING
FIRE DEP'T

(embossed on wind break)

DIETZ (logo)

Patent Information: (stamped on wind break)

PATENTED
AUG. 27, 07

Remarks:

The Dietz Fire Department lantern was often seen attached to, or hanging from, the pumper, chemical, and ladder wagons. Later, when fire wagons were motorized, the very same lanterns fit special lantern holders on the trucks. The globes were usually clear but could be ordered with a ruby globe for use as a safety light.

The fount of all fire lanterns and many railroad signal lanterns are filled with cotton waste as a safety measure. The cotton absorbs the kerosene and prevents spills. If a lantern is upset, the fuel can only drip out slowly.

Variations:

The American LaFrance Fire Equipment Company of Elmra, N.Y. was a major producer of fire fighting apparatus and special ordered lanterns with their name on the draft guard. Fire lanterns are highly polished brass, nickel, plated brass, or steel plated with nickel. Lanterns with a copper fount were only available until 1914.

From 1881 to 1887, Dietz made the New York Fire Department Tubular lantern without a wind break. The hanger and wind break first appeared on the Dietz Tubular Fire Department lantern in 1904.

Plate 7.1
Dietz King Fire Department, 1897-1940, 14.5 in.(36.8 cm) 5.75 in.(14.6 cm) $220-$310 USD

The Queen Brass Fire Tubular (1901-1910) had square tubes, a wind break, and a hanger. The 1917 Dietz catalogue shows the No. 2 "Wizard" Fire Department Lantern which comes in red enamel, polished brass, and nickel plate. The Wizard uses a short globe.

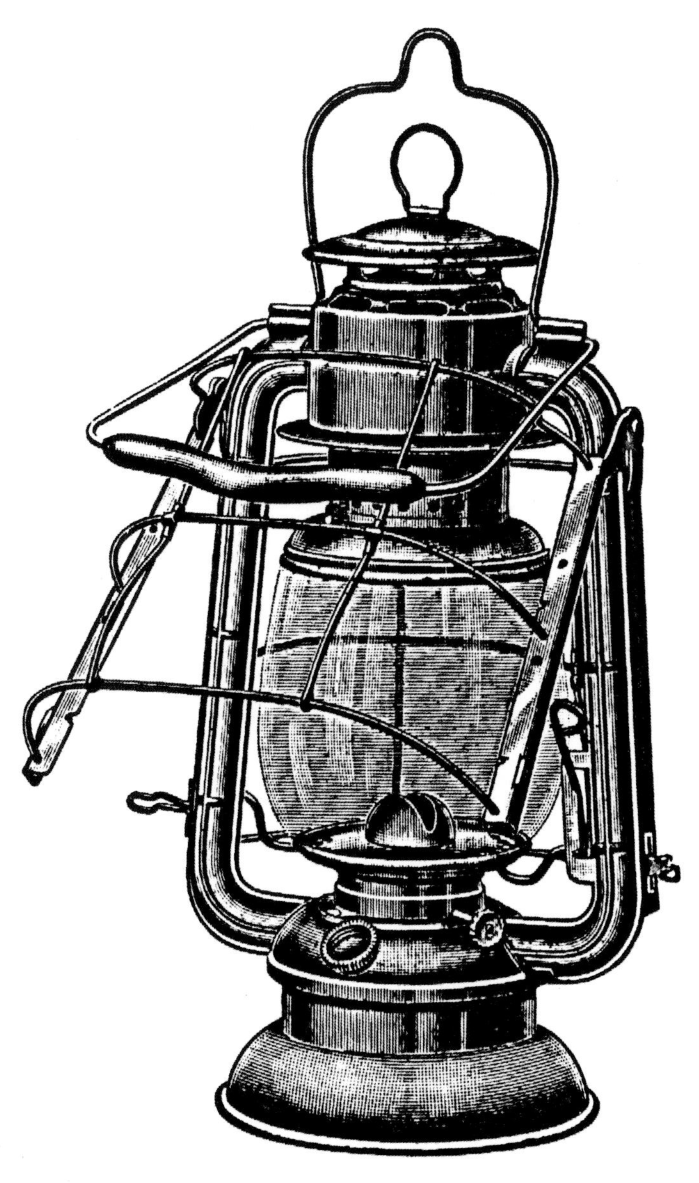

Plate 7.1a
The No. 2 Wizard is a cold blast fire lantern with a D-Lite globe.

Dietz Buckeye Dash

Plate 7.2 **Description**:

The Dietz Buckeye Dash is a hot blast steel lantern with brass fittings. The burner is a brass No. 1 with a 0.625 inch (1.6 cm) wick. The back of the lantern has a tin reflector with a large spring loaded clip that can open to about 0.5 inch (1.27 cm). The correct globe for this lantern is the early, pear shaped, DIETZ No. 0 TUBULAR. Black or blue are color choices.

Markings: (embossed on fount)
DIETZ N.Y. U.S.A.
(embossed on reflector)
DIETZ BUCKEYE DASH
(embossed on globe)
DIETZ No. 0 TUBULAR

Patent Information: (on center tube)
PATENTED

Apr	??	-0
Aug	??	90
?	??	91
Apr	28	91
May	21	93
May	23	98
July	11	99
July	26	01
Mar	13	06

Remarks:

Here is what the 1902 Sears Roebuck Catalogue says about the Buckeye:

> This is really a very handy combination. It serves as a hand lantern and a side or dash lamp. We furnish it

Plate 7.2
Dietz Buckeye Dash, 1898-1946, 13.25 in.(33.7 cm) 6.0 in.(15.2 cm) $220-$310 USD

japanned blue. The lamp can be fastened under the body of a vehicle by means of a bracket. We furnish this lamp with our new bull's eye lens - a bull's eye attached to the globe plate. It is superior in every way to the bull's eye globe. No. 1 burner, 0.625 inch (1.6 cm) wick, No. 0 globe. Weight about 2 1/4 lbs. Each...70c

The Dietz Buckeye Dash Lamp, as its name implies, is a headlight that clips to the dashboard of a buckboard or buggy. The bull's eye is just a magnifying lens designed to focus the light on the road.

A dash lantern would usually have a bull's eye globe, with the lens molded onto the glass as the globe is made. Because the Buckeye Dash lantern has the bull's eye attached to the globe plate, the more expensive bull's eye globe is not needed.

Note that a dash lantern does not have a red lens in the reflector as it would not show when clipped to a dashboard. A wagon lantern (Plate 7.16) is similar to a dash lantern but mounts at the side so the ruby "tail light" lens is visible.

Variations:

An earlier model Dietz No. 13 Tubular Side and Dash lantern has a flat fount. The No. 13 was only made from 1886 to 1897.

Plate 7.2a
This is the rear view of a dash lantern showing the dash spring and no red lens.

C. T. Ham No. 2 Hot Blast

Plate 7.3 **Description:**
The hot blast version of the C. T. Ham No. 2 is a steel and brass lantern with an unusual domed hot air chamber under the burner. This No. 2 has a brass burner and small fill cap. It needs the large fount because of the large No. 2, domed burner and 0.875 inch (2.2 cm) wick. The Ham No. 2 has an outside lift and an early example of a crimped bottom. The majority of the lantern is hand solder construction and the correct globe is a HAM No. 0. According to Ham advertising the lantern is constructed of IX tin.

Markings: (embossed on fount)
C. T. HAM MFG. CO. No 2
(embossed on globe)
HAM
3
(on burner dome)
C. T. HAM MFG. CO. U.S.A.
HOT BLAST NO. 2 CLIPPER & S.S.S.

Patent Information:

Ham usually put patent dates on the front of the burner air chamber just below the burner, but this lantern has none.

Plate 7.3
Ham No. 2 (hot blast version), 1898-1914, 14.0 in.(35.6 cm) 6.5 in.(16.5 cm) $100-$150 USD

Remarks:

The No. 2 is a simple lantern with no fancy frills.

The C. T. Ham Manufacturing Company began in 1886, in Syracuse, New York. The company made hot and cold blast hand lanterns, railroad style signal lanterns in steel and brass, a No. 39 signal lantern, non-removable globe models, and a lantern called the Empire No. 0 Hood Car Inspector. The S. S. S. on the burner dome refers to another Ham lantern called the "Side Spring Safety." The No. 2 Side Spring Safety lacks a side lift but has spring wires on the tubes to hold the globe up or down. Clipper is another name used by C. T. Ham on many other lanterns.

The correct globe is a Ham's No. 0 but these are difficult to find. More Ham's lanterns have survived than Ham's globes. This HAM globe has a crack. Care must be taken when lighting these large burners as setting the flame too high on a cold globe will crack it. An unmarked, old style, fluted globe would also be appropriate.

Variations:

Advertisements for Ham lanterns give the choice of finish as plain (tin), nickel plate, solid brass or japanned. The japanned finish could be blue or black. Compare this No. 2 burner lantern to the smaller Ham Clipper in Plate 6.11 and to the cold blast version of Plate 6.10.

Another model in the Ham line is this lantern with a reflector added to make a large dash lantern.

Plate 7.3a

Adlake Switch Stand Lantern

Plate 7.4 **Description:**

This dead flame lantern has a nickel plated, brass burner with a 0.875 inch (2.2 cm) flat wick. The balance of the lantern is crimped, soldered and riveted steel. This switch lantern has two red and two green colored, 4 inch (10.2 cm) lenses. The fount drops into the lantern through the hinged top. The top has a hasp for locking the light closed. One corner has a 0.75 inch (1.9 cm), round flame viewing window. Original finish appears to be black paint over the steel. This model has two vertical tubes that mount on a fork attached to the top of the switch stand.

Markings: (embossed on the crown)

THE NON-SWEATING
ADLAKE LAMP
CHICAGO

(embossed on the pot)

THE ADAMS & WESTLAKE CO.
CHICAGO NEW YORK PHILADELPHIA
FOR KEROSENE ONLY

Patent Information: (on brass side plate)

PATD IN US JAN 24 1899 JUN 12 1900
JAN 13 1903 AUG 4 1903 OCT 27 1903
MAR 29 1904 AUG 13 1907
OTHER PATS PEND
THE ADAMS & WESTLAKE CO
CHICAGO NEW YORK PHILADELPHIA

Plate 7.4
Adlake Switch Stand, 1899-1940, 14.75 in.(37.5 cm) 7.0 in.(17.8 cm) $270-$360 USD

Remarks:

This Adlake railroad switch stand lantern uses colored lenses to indicate the position of the railroad track switch. When the lantern is turned so the green lights face up and down the main track, the switch is aligned for that track. When the switch is thrown to the side track, the lantern rotates so the red lens shows down the main track.

This Adlake model 169 has mounting tubes designed to slip over round posts on the fork. The fork has a square socket that fits the square top of the switch stand. The post and lantern turns 90° when the switch is moved from one track to the other. Compare to the Adlake Semaphore lantern of Plate 7.21.

Variations:

There are different color lenses for switch stands used in yards or derail switches. A derail is a short track set to drop a railroad car on the ground. Why? It is a safety measure to prevent a car set on a siding from rolling onto the main line, where it would be a hazard.

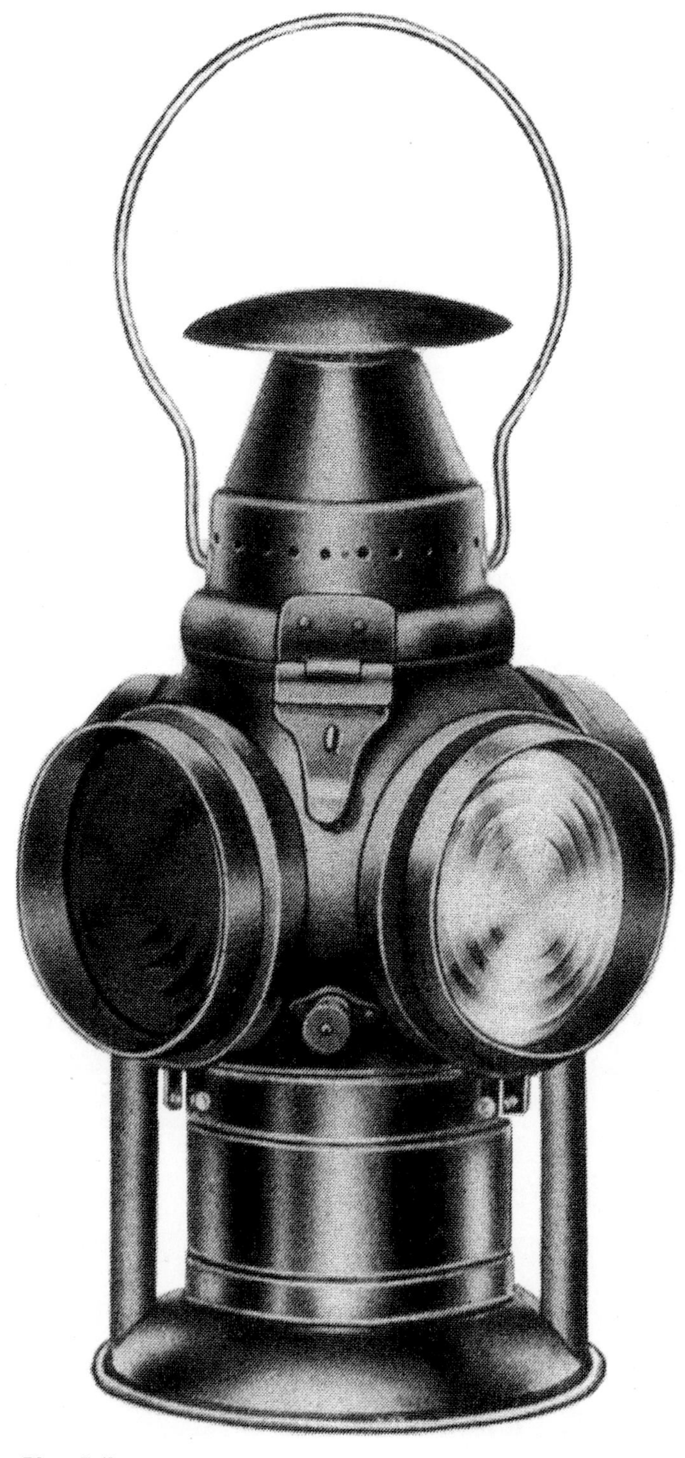

Plate 7.4b
Cut of the model 169 from the Adlake catalogue.

Plate 7.4a
This Plate shows the Adlake lantern in position on its switch stand.

Dietz Hy-Lo

Plate 7.5 **Description**:

The economy Dietz Hy-Lo model is an all steel, hot blast, tubular lantern. The Hy-Lo has a large steel fount cap and steel No. 1 burner with a wire ring adjust. The Hy-Lo has no lift, just a small loop of wire on the globe plate. The construction is mostly machine crimp with some hand soldered parts. All were made in New York City or Syracuse, New York, U.S.A.

Markings: (embossed on crown)
DIETZ (logo)
(embossed on fount)
DIETZ HY-LO
NEW YORK U.S.A.
(one of many variations)
DIETZ HY-LO
CITY OF LA
(embossed on cap)
DIETZ (logo)
(on globe)
4 H
LOC - NOB
RECD IN U. S. A.
DIETZ (logo)
FITZALL
N. Y. U. S. A.

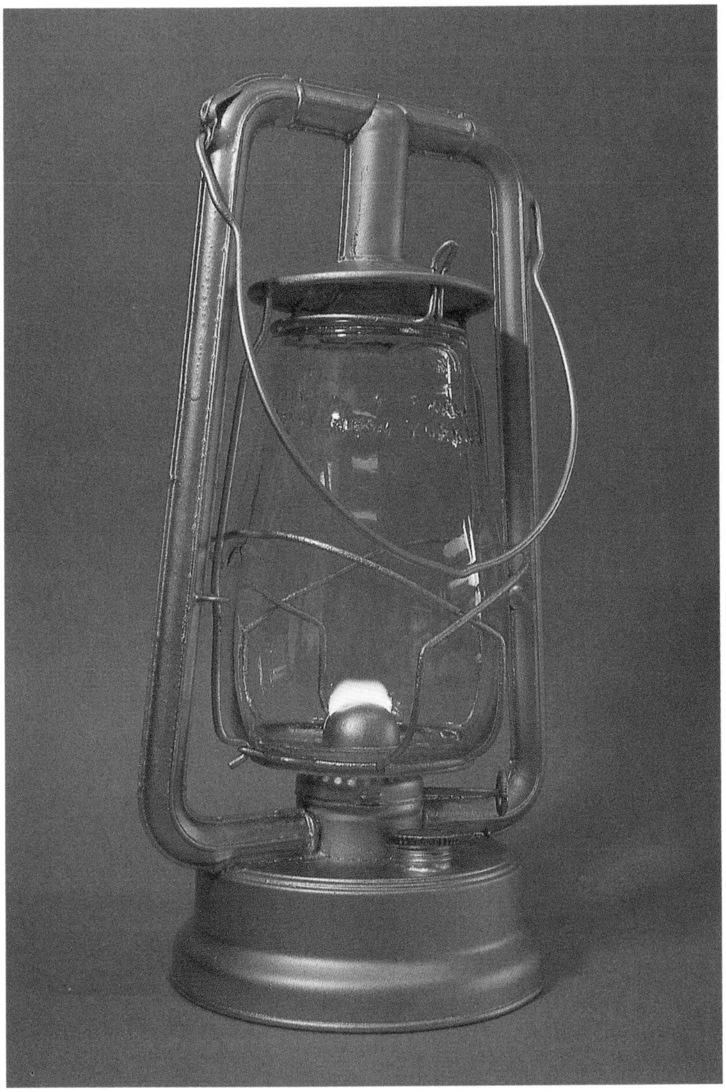

Patent Information: (on center tube)
PATENTED
MAY- 24 -20
S- 11 -29
(and later)
PATENTED
MAY- 24 -20
S- 12 -41

Remarks:

The Dietz Hy-Lo is one of the more readily available antique lanterns because of its popularity. The cost of the Hi-Lo was kept to a minimum by using simple construction and eliminating the cam type lift. Because of its popularity, watch for the Hy-Lo in your favorite western movies including "The Treasure of the Sierra Madre" staring Humphrey Bogart.

This lantern was available is japanned black or tin plated. This specimen has minor rust and dents. Note the patent dates stamped into the center tube above the globe.

Since the Hy-Lo has a thumb lift on the globe plate it has always had a unique globe guard. The guard was improved for better globe holding force about 1940 as shown in Plate 7.5a.

Variations:

Because it was inexpensive, the Hy-Lo was popular among municipalities and contractors. One could expect to find the Hy-Lo embossed with the names of state, county, or city governments. This lantern is commonly found with clear or ruby LOC-NOB globes.

Finish and color also varied for each special order. Red was popular and tin plate was common. The 1899 to 1919 Hy-Lo has the same lines but with a small fill, brass burner, and old style globe. Compare this lantern to the 1913 Monarch of Plate 7.19.

Plate 7.5
Dietz Hy-Lo, 1899-1945, 13.5 in.(34.3 cm) 6.0 in.(15.2 cm) $20-$50 USD

Plate 7.5a
The pre 1940 Hy-Lo globe guard is shown on the left with the improved guard.

QUALITY TAG

Is attached to each GENUINE DIETZ LANTERN as a guarantee of QUALITY and PERFECTION. All DIETZ LANTERNS are carefully inspected. You take no risk when buying a DIETZ make as the DIETZ reputation stands back of each LANTERN. DIETZ makes a LANTERN for Every Purpose; be sure this is the right one for your use.

R. E. DIETZ COMPANY

Largest Makers of Lanterns in the World

Established 1840

New York, U. S. A.

(over)

Plate 7.5b
The label that came attached to every Dietz lantern.

Buggy/Auto Side Lantern

Plate 7.6 **Description:**
This Ford Model T Side Lamp is made of black painted steel with brass trim. It is a cold blast lantern with a steel, No. 1 burner that uses a 0.625 inch (1.6 cm) wick. Two sides have silvered reflectors and the two glass plates are beveled. One glass side is hinged for access to the wick. The burner is part of the fount that removes with a twist from the bottom. The construction of this side lantern is mostly hand soldered parts.

Markings: (embossed on crown)
Ford (script)
JNO. W. Brown MFG Co.
(or) The Victor Lamp Co.
MODEL 110 (or MODEL-2)
COLUMBUS. O. (or CINTI-O)

Remarks:

All kerosene side lamps are actually small lanterns. Most use a 0.625 inch (1.6 cm) wick. Cut glass panes are typical. Two of these lanterns were used on about half of the 15,458,781 Model T Fords built. The Model T made up about half of the cars produced up to 1916. This works out to about 15 million of these buggy and auto side lamps made!

There are numerous styles with square being the older and rounded being the newer style. Manufacturers offered a variety of finishes including: painted steel, nickel plate, polished brass, or any combination. Each style had a different name. Dietz made the Octo Driving Lantern, The Ideal, The Orient, The Regal, The Union, on and on. Some have colored lenses or jewels on the rear, side, or both. Dietz had one called The Sterling Junior Side Lamp that was very fancy.

A very similar kerosene lantern is used as the tail light on early cars. One side has a round red lens and the other three sides are tin. Kerosene headlights were used on the earliest automobiles but the light was not bright enough for higher auto speeds. Gas lights were the popular solution. Acetylene gas was generated from Calcium Carbide in a tank on the running board. The gas passes through rubber tubing to burners in the headlight(s). An acetylene flame is bright, white, and set against a parabolic reflector, throws a good distance.

Imagine taking a nighttime drive in a 1910 Model T Ford. You fill and light the single kerosene tail light and two side lights. Hopefully it's not windy. Put calcium carbide and water in the gas generator, then adjust the drip. If the generator is not clean the old residue should be removed first. Once you start making gas the process cannot be stopped until the carbide is used up. Now you can light the burners in the headlights. Don't forget, the reason these cars have gas and kerosene lights is they have no low voltage electrical system. The engine is started with the hand crank.

Plate 7.6
Buggy/Auto Side Lamp, 1900-1916, 11.75 in, 29.8 cm) 5.5 in.(14.0 cm) $120-$170 USD

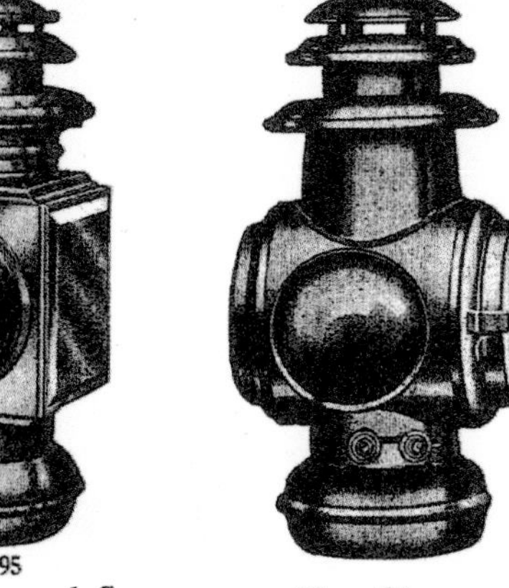

Plate 7.6a
Side and tail lanterns for buggies and early autos.

Plate 7.6b
As styles changed the side and tail lights shapes became rounder.

Dietz ACME Inspector Lantern

Plate 7.7 **Description**:

The Dietz ACME Inspector is an all steel lantern with a hood that is 6 inches deep, fitted with a 5 inch, silvered glass reflector. The ACME has a No. 1, long cone burner, 0.625 inch (1.6 cm) wick, and uses a LOC-NOB globe. This specimen has a large fount cap, an overall tin finish, and top lift. The gray finish, straight handle and LOC-NOB globe date this lantern between 1940 and 1954. From the collection of Scott E. Schifer.

Markings: (on back of reflector)

DIETZ ACME INSPECTOR LAMP
NEW YORK, U.S.A.

(on fount)

DIETZ NEW YORK, U.S.A.

Patent Information: (on center tube)

PAT.

JULY-	11	-9?
JULY-	26	-04
MAR.-	13	-06
???-	4	-0?
MAR-	8	-10
AUG-	13	-12
SEPT.-	16	-13
SEPT.-	9	-13
S-	2	-27

Remarks:

Popular with railroads, the Inspector was an expensive general purpose lantern. The correct globe is a DIETZ No. 0 TUBULAR in Inspectors with a curved handle, and a DIETZ LOC-NOB globe in Inspectors with the straight handle.

I was told that this lantern was used on a caboose of the Tonopah-Tidewater Railroad. The Tonopah-Tidewater RR, operated from Oct. 1909 to June 14, 1940, on 250 miles of track in Southern California.

This is the kind of story that adds nothing to the value of a lantern as it cannot be confirmed. Railroads would often have their name embossed on the lantern handle by the manufacturer. However, Inspectors were purchased by railroads and non-railroad customers alike.

Variations:

Variations include a curved rear handle before 1914. An optional globe guard was available beginning in the 1920s. Finish was tin coated until 1940 then Dietz painted the Inspector in gray, blue-gray, and then blue enamel paint. The Dietz Protector Track walker's Lantern is a similar model with a rear red lens made from 1913 to 1945.

There are two versions with dash springs, the Dietz No. 0 Tubular Reflector (1888-1905) and the Dietz Beacon Dash made from 1905 to 1925. Adlake sold the Dietz ACME as the No. 114 Inspector's lantern.

Plate 7.7a
The rear handle is curved and the fill cap is small on Inspectors made before 1914.

Plate 7.7
Dietz ACME Inspector, 1900-1954, 14.75 in.(37.5 cm) 6.5 in.(16.5 cm) $60-$70 USD

Hurwood Aladdin

Plate 7.8 **Description**:

The Hurwood Aladdin is a large, steel, cold blast lantern with a small, one piece, brass fount cap, and No. 2, brass burner with a 0.875 inch (2.2 cm) wick. This lantern has a unique outside lift that swings the globe off the burner and out of the lantern. The original finish is bright tin or japanned black. The age and type of lantern indicates that an early, pear shaped, generic No. 0 globe is appropriate.

Markings: (embossed on fount)

THE HURWOOD MFG. Co.
ALADDIN

Patent Information: (none)

Remarks:

The Hurwood Manufacturing Company was started in 1903 by John A. Hurley in Bridgeport, Connecticut. This lantern is an example of a major restoration. When found, the fount of this lantern had extensive rust damage. After professional rust stripping in a Caustic Soda dip, the lantern was primed. The holes were much too large for a simple application of gas tank sealer so auto body tools and techniques were used to fill the voids. An application of polyester auto body filler is used to replace and reproduce the shape of the missing metal. The filler is filed to shape and sanded smooth. Next comes the fount sealer and final color coat of paint. Detailed lantern restoration techniques can be found in Chapter 10.

Normally a lantern would not warrant such extensive restoration as the value of the lantern will never offset the cost. In this case, however, the lantern has unique features that give it historic value. The Hurwood Aladdin has a unique and complex lift design patented by Kurt Ludwig Stendahl on 11-18-07. As the globe lifts, a rod raises the chimney off the globe. Once the globe plate has cleared the burner dome, the entire globe swings over the dome and out of the lantern. As a globe settles behind the burner, the chimney drops back down to hold the globe in place. The burner is clear for lighting and the globe is free for removal. There is no need for a lift ring on the crown.

Plate 7.8
Hurwood Aladdin, 1907-1915, 15.0 in.(38.1 cm) 6.75 in.(17.0 cm) $100-$150 USD

The Aladdin has the easiest access to the burner of any lantern and yet it was a complete failure in the market. It may have failed because the complex lift added more cost than the market would bare. This just goes to prove that just because something is designed better, it is not guaranteed to be successful.

Lanterns made by small companies are usually simple, common and not very interesting because they follow the trends rather than push the state-of-the-art. The Aladdin is an exception to this rule.

Plate 7.8a
The Hurwood lift was complex and not very successful. Here is the Hurwood lift on the rare hot blast Aladdin.

Plate 7.8b
The Hurwood before restoration.

Dietz Standard Brakemen's Lantern

Plate 7.9 **Description**:

The Dietz No. 39 Standard is a steel, kerosene, brakemen's lantern with a bell bottom (solid base) and single wire guard. The slip out pot has a No. 1, brass burner with a 0.625 inch (1.6 cm) wick. The lantern has an original Dietz ruby, No. 39 globe. First made in New York City and then in Syracuse, New York.

Markings: (embossed on globe)
DIETZ (logo)
NEW YORK U.S.A.
(embossed on fount)
* No. 39 STANDARD *
DIETZ NEW YORK U.S.A.

Patent Information: (none)

Remarks:

The Dietz No. 39 Standard is just one of a whole series of lanterns to use the No. 39 globe. The Standard is typical of the railroad style signal lantern used by the brakemen to signal the conductor and engineer. The No. 39 was introduced in 1887 and went through several redesigns. Around 1900, the name "Standard" was added and it continued to be made until 1944. This so called "standard" lantern came with a dizzying variety of options. It could be ordered to burn kerosene, lard oil, or signal oil. The pot type can be bayonet catch, side spring, drop in made of tin, or glass. It can have a bell bottom or wire bottom, and be made of tinned steel, brass, or nickel plated brass. It can be found with a single or double wire guard and any of the five standard globe colors.

In the early days of railroading there were no automatic air brakes so the job of the brakemen were to set and release the hand brakes on the freight cars. The conductor generally performed his task from the caboose and he had to signal the engineer and brakemen to start, stop, reverse, and so forth. The brakemen, spaced along the length of the train, often had to relay the conductor's signals to the engineer in the locomotive. At night, railroad lanterns were used so the hand signals could still be seen. The engineer and brakemen also carried these lanterns for light as well as signaling.

Signal oil was the common fuel before World War 1 and the burners are a little different. Refer to the description of fuels and burners in Chapter 3.

Plate 7.9
Dietz No. 39 Standard, 1900-1944, 9.75 in.(24.8 cm) 6.0 in.(15.2 cm) $70-$80 USD

Nearly all lantern manufacturers made railroad style lanterns. The design and technology was public domain plus loss and breakage was prevalent on all railroads.

This specimen has no trace of paint, minor dents, and very little rust. It appears the original finish was terne plate rather than tinned steel. This may indicate it was made during World War Two when defense needs reduced the amount of tin available.

Variations:

The No. 39 in similar to the Dietz Steel Clad, the Vulcan, and to the Dietz No. 6 railroad lanterns. Compare to the older Handlan-Buck of Plate 7.10 and the Adlake Reliable of Plate 7.20.

Handlan-Buck "The Handlan"

Plate 7.10 **Description**:

"The Handlan" is wire bottom, double wire guard, dead flame, steel, brakemen's lantern. It has a 6 ounce, twist off pot with an internal adjust, brass, signal oil burner. The wick is 0.625 inch (1.6 cm) and the construction is by solder and rivet. This Handlan takes a No. 39 or No. 223 globe and has no railroad markings. "The Handlan's" original finish appears to be zinc or tin dip.

Plate 7.10
Handlan-Buck, The Handlan, 1901-1918, 10.5 in.(26.7 cm) 7.0 in.(17.8 cm) $80-$90 USD

Markings: (on crown)
"THE HANDLAN"
HANDLAN - BUCK MFG. Co. ST. LOUIS

Patent Information: (on crown)
Patented
Dec. 18, 88.
Jun. 18, 89.
Sep. 24, 89

Remarks:

In 1869, the golden spike at Promontory Point, Utah, signaled the completion of the Transcontinental Railroad. The same year, Alexander H. Handlan became an employee of Myron M. Buck's railroad supply manufacturing business in St. Louis, Mo. This was the right business, in the right place, at the right time. By 1875, M. M. Buck & Company made two dozen different fixed and removable globe signal lanterns plus tons of railroad car fixtures.

By 1901, Mr. Handlan had bought out Mr. Buck and changed the company name to Handlan-Buck Manufacturing Co. A. H. Handlan, Jr. took over the company when A. H. Handlan, Sr. died on May 28, 1921. The Handlan Company celebrated their 100 Anniversary in 1956. The company continued to sell lantern parts well into the 1970s.

"The Handlan" is a fancy name that appears to do honor to the company president, A. H. Handlan. This classic, dead flame lantern could have a number of uses on a railroad. Fitted with a ruby globe this lantern would be used to signal the engineer to "High Ball" or proceed. With an amber globe it could be used as a caution marker for a track crew. A blue globe was used to mark equipment that was having work done, and a clear globe was used for signaling and light.

Standard Railroad Signals

Vertical up and down: **Proceed**
Horizontal side to side: **Stop**
Swing in a circle: **Back up**
Held above the head: **Release Brakes**
Horizontal swing above head: **Apply Brakes**

"The Handlan" has solder construction and does not appear in the 1918 catalogue that features lanterns that are electrically welded and "Constructed Entirely Without the Use of Solder." Railroad style lanterns were the first to eliminate hand soldering.

Compare to the later Adlake Reliable brakemen's lantern of Plate 7.20 and the older Universal Conductor's lantern of Plate 6.4.

Variations:

There is an identical lantern named "The Buck" and both had an optional fiber handle for electric railroad use. Another variation of "The Handlan" has a bracket to lock the handle up. The burner can be No. 1, No. 1 Ex. Beacon, or Convex.

Dietz Vesta

Plate 7.11 **Description:**

The Dietz Vesta is a general purpose, tubular, cold blast, all steel, kerosene lantern. The most striking features are the cold blast tubes that feed fresh air to the flame. The Vesta has an open base, single wire guard and a twist off fount. The steel, No. 0, domed burner uses a 0.38 inch (1.0 cm) wick and has a large, outside adjust. The construction is mostly machine crimp with electrical welds and some hand soldered parts. The Vesta's finish is terne plate. The Vesta produces 4 candle power and burns 16 hours on one fill. The patent dates indicate this short style Vesta was made after 1943. It has no owner markings.

Markings: (embossed on crown)
NEW YORK ** U.S.A. **
DIETZ VESTA (in oval)
(embossed on globe)
DIETZ (logo) VESTA
NEW YORK U.S.A.
CNX (Corning logo)
(embossed on fount)
KEROSENE OIL - TAKES DIETZ VESTA GLOBE

Patent Information: (on brim)
PATENTED
JULY 30-07
MAY 4-09
JUNE 1-09
DEC 18-10
S - 7 - 26
(alternate)
PATENTED
NOV 8 '27
APR 3 '28
DEC 25 '28
MAY 4-20
S-10-42 (or) S - 4 - 43

Remarks:

The Vesta became Dietz's answer to Adlake's popular Kero of Plate 8.1. Advertised as giving three times the light for one third the fuel, however the Vesta did cost more to buy.

The Vesta lantern is special because it is a cold blast railroad style lantern. Since the first use of signal oil, railroad lanterns have been the dead flame type. The Vesta was a major improvement in light output over the traditional railroad style. The Vesta is named after the Roman Goddess of hearth.

Variations:

There were many variations of the Dietz Vesta. Designed for kerosene, the Vesta started with a No. 39 globe in 1896, but was changed to take a Junior globe in 1902. Apparently this did not satisfy the designers as an entirely new and smaller globe was developed. The Vesta globe, at 4.25 inches (10.8 cm), is an inch (2.54 cm) shorter than the No. 39 globe. The volume and draft of the Vesta globe is optimized for maximum brightness from the small 0.38 inch (.97 cm) wick.

Vestas before 1907 have a closed bottom rather than a wire bottom. In 1924 the Vesta's height was reduced about an inch (2.54 cm). The older Vesta has a taller bell and fount. All colors of globes were used in this popular lantern. The different finishes include bright tin, olive green, and solid brass. The Vesta can be found with a fiber bail, mounting bracket, heavy base, or blinders. The Vesta's rugged construction, small size, economy, and bright flame made it popular for ships, railroads, the U.S. Army, as well as for general use.

Plate 7.11
Dietz Vesta, 1906-1950, 10.0 in.(25.4 cm) 6.5 in.(16.5 cm) $60-$260 USD

RAILROAD

Dietz New Vesta

Complete with No. 500 Burner, ½-inch Cosmos Wick and Vesta Globe; lighted and regulated from the outside; made on the tubular principal. For all railroad service.

New Vesta Railroad, for coal oil.

Per dozen **$40.00**

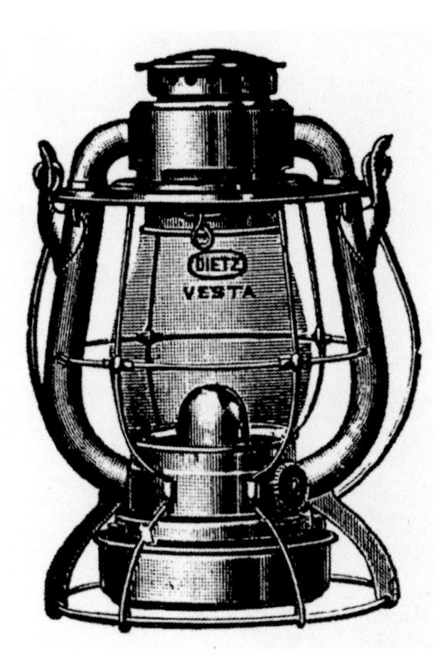

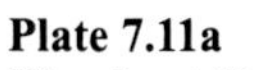

Plate 7.11a
The short Vesta lantern was rather pricey at $3.32 each.

Dietz No. 30 Beacon

Plate 7.12 **Description:**

The Dietz Beacon is an all steel, cold blast, side lantern with a single air tube at the back. The Beacon uses a No. 0 globe, unusual front lift, a No. 2, domed burner, with a 0.875 inch (2.22 cm) wick. It has a hood that is 12 inches diameter (30.5 cm) and 7 inches (17.8 cm) deep. The hinged, flip top fill cap is unique and needed due to the tight location. The Beacon has a square, machine crimp bottom and many soldered parts. The Beacon has a bail and a wire loop for hanging.

Markings: (embossed on bell)

DIETZ 30 NEW YORK U.S.A.

(embossed on globe)

DIETZ (logo) FITZALL
N. Y. U. S. A.
H18 LOC-NOB
PAT'D 12-4-23

(embossed at bottom front of the reflector)

DIETZ 30 BEACON LIGHT
NEW YORK U. S. A.

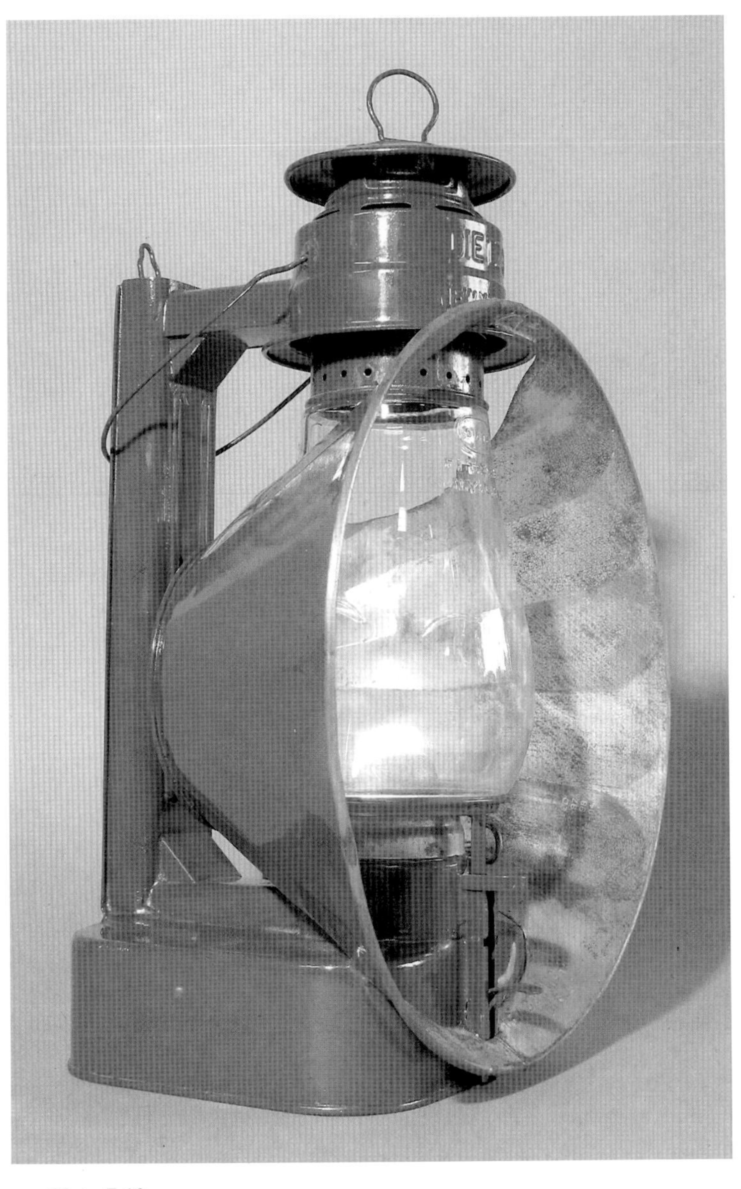

Plate 7.12
Dietz Beacon, 1906-1945, 15.0 in.(38.1 cm) 6.0 in.(15.2 cm) $60-$70 USD

Patent Information: (none)

Remarks:

The tubular side lantern was an important innovation in the development of hot blast technology. Using the structure of the lantern to force air into the burner was the invention of John Irwin.

This Dietz No. 30 Beacon was made between 1913 and 1945. It can be identified by a hood that is 12 inches (30.5 cm) diameter, 7 inches (17.8 cm) deep and its No. 2 burner.

Before the cold blast Dietz Beacon there was a series of hot blast side lanterns that used a single large tube at the rear. There is a No. 15 Dietz Tubular Side lantern made from 1873 until 1906. The No. 15 has no hood and uses a No. 1 burner. The same lantern with a No. 2 burner was called the Dietz No. 17 Tubular Side lantern. A 6 inch (15.2 cm) mirrored glass reflector was added to both sizes in 1875.

In 1894, the cold blast predecessor to the No. 30 was the Dietz No. 25 Tubular Side that also has a 6 inch (15.2 cm) mirrored glass reflector.

The No. 30 is finished in japanned blue and a bull's eye globe was an option.

Variations:

The No. 60 Beacon is a large version of the No. 30. The No. 60 has a 1.5 inch (3.8 cm) wick and the hood is 16 inches (40.6 cm) in diameter and 10 inches (25.4 cm) deep. The No. 60 was only made from 1913 to 1930.

Plate 7.12a
A close-up of the globe lift and reflector.

Plate 7.12b
The No. 30 (front) is shown with the No. 60. More about this No. 60 can be found at www.classiclantern.com/sixty.htm

Dietz Champion

Plate 7.13 **Description:**

The Champion is an all steel, cold blast, motor truck side lantern. It has a 5 inch (12.7 cm) clear lens and no rear lens. The brass insert pot has a brass, domed burner that uses a 0.5 inch (1.26 cm) wick. The reflector is made of polished aluminum. Machine crimp construction and hand solder is used. The finish is black enamel paint.

Markings: (embossed on back)
DIETZ (logo)
* CHAMPION STEEL LAMP *
JAN. 22 -07 & PATENT APPLIED FOR
(on lens)
DIETZ

Patent Information: (embossed on back)
JAN. 22 -07 & PATENT APPLIED FOR

Remarks:

The Champion is the largest and brightest side lantern in the Dietz line. The cold blast air circulation is between the housing and the aluminum reflector.
The Champion has a clamp on each side for mounting on either side of the truck. It is one of many lantern models discontinued in 1930 when the New York City factory was closed.

Variations:

The Champion was made as a side light for 2, 3, or 5 ton trucks. A slightly smaller side lamp is the Royal Junior for half-, 1-, and 2-ton trucks. For size comparison refer to the Ford Model T side lantern shown in Plate 7.6. An even smaller lantern is the Dietz Eureka of Plate 7.27.

Plate 7.13
Dietz Champion, 1907-1930, 11.5 in.(29.2 cm) 7.75 in.(19.7 cm) $140-$200 USD

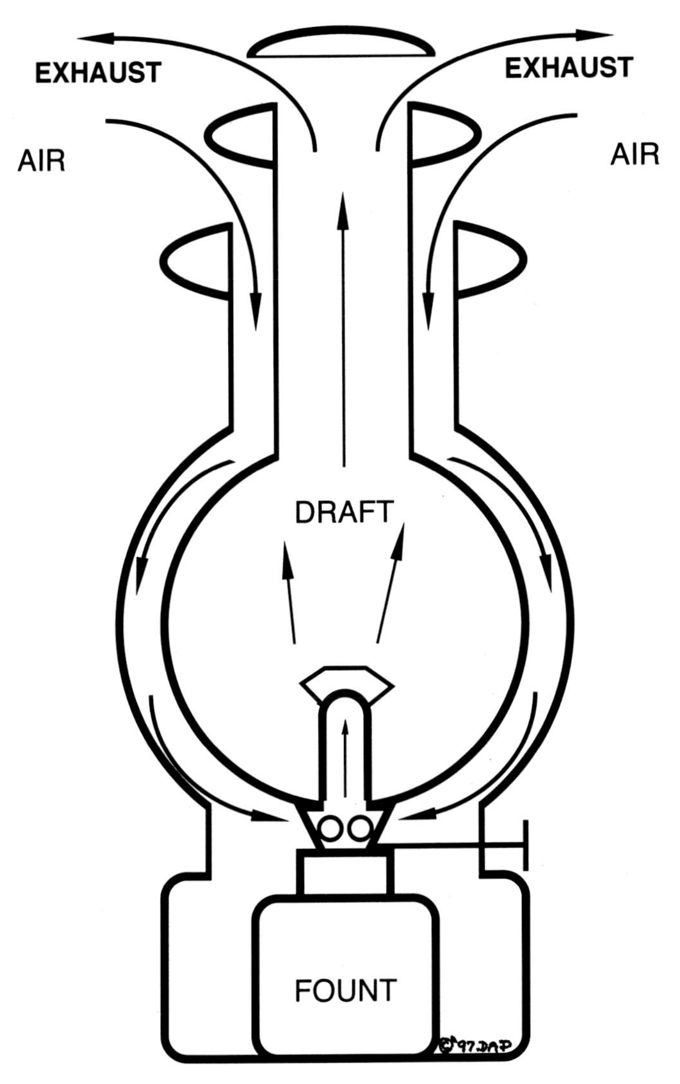

Plate 7.13a
Diagram showing the cold blast air circulation of a typical auto lantern.

Embury No. 2 Camlox

Plate 7.14 **Description:**

The Embury Camlox is a cold blast lantern made mostly of steel with a copper fount. It has a two piece, brass fill cap and a No. 2, brass, domed burner with a 0.875 inch (2.2 cm) wick. The Camlox has an inside lift and the steel parts are painted black. The globe type is a No. 0 and the bell has some unusual embossed decoration. The large fount holds one pint of kerosene.

Plate 7.14
Embury Camlox, ca. 1911, 15.0 in.(38.1 cm) 6.75 in.(17.0 cm) $80-$100 USD

Plate 7.13b
The front swings open to allow lighting and removal of the burner.

Markings: (on fount)

No. 2 C. B. CAMLOX EMBURY MFG. CO. WARSAW, N.Y. U.S.A.

Patent Information: (none)

Remarks:

The Embury Manufacturing Company of Rochester, New York was incorporated November 27, 1908 by William C. Embury. The company designed and produced a variety of tubular, dead flame and railroad style lanterns. The factory moved to Warsaw, New York in 1911, where it remained until it was closed in 1952.

This lantern is embossed Warsaw, N.Y. so it was made after the move from Rochester. The C. B. on the fount refers to the "Cold Blast" draft technology used on this lantern. The cold blast patent provided a measurably brighter light and was well worth advertising. The No. 2 on the fount refers to the larger, 0.875 inch (advertised as 1 inch) wick used in the burner.

Seldom will the collector be so lucky as to have the original owner put his name and date on a lantern.

Embury built a full line of farm and contractor's lanterns, offering clear or ruby globes. This lantern was one of their more unusual offerings has it had a very heavy, solid copper fount. The large fount fill was a new feature in 1913. Note the decorative embossing on the cold blast bell. In 1913, the copper fount Camlox price was $15.00 per dozen or $12.00 for the tin fount.

Variations:

The Camlox was also available in a japanned blue dash version with a reflector and bull's eye lens for $15.00 per dozen. Embury also made a variety of dead flame kerosene and battery powered railroad style lanterns. The Embury No. 140 Railway Inspectors Lantern was almost identical to the Dietz ACME Inspector of Plate 7.7.

Plate 7.14a
Close up of the copper fount shows the lantern was not made in Rochester but at the new Warsaw factory.

Scratched on the cap

C. E. D.
1914

Scratched on the bottom

C. E. Dirtruk

Plate 7.14b
The owner of this lantern was rightly proud of his purchase and scratched his name into the bottom and his initials in the brass fount cap.

Dietz No. 2 Blizzard

Plate 7.15 **Description:**

This all steel, cold blast lantern has a large, two piece fount fill cap embossed with the Dietz logo. It has a LOC-NOB globe, inside lift, all steel No. 2, domed burner, with a 0.875 inch (2.2 cm) wick and bent wire wick adjust. The air tubes are machine embossed for strength. Machine crimp construction and hand solder is used. The Blizzard is advertised as giving 10 candle power for 20 hours.

Markings: (embossed on fount)

DIETZ N. Y. U.S.A.
NO. 2 BLIZZARD

(embossed on crown)

DIETZ (logo)

(embossed on globe)

H8 LOC - NOB
PATD. 3 - 10 - 14
BLIZZARD
DIETZ (logo) FITZALL
N. Y. U. S. A.

(full text fill cap)

DIETZ (logo)
MADE IN THE UNITED STATES OF AMERICA

Patent Information:

PATENTS

FEB-	1	-98
AUG-	1	-99
AUG-	7	-00
JUL-	26	-04
AUG-	16	-06
NOV-	19	-07
MAY-	8	-10
JUN-	21	-10
MAY-	7	-12
M-	7	-27

Remarks:

In production from 1912 to 1945, the Blizzard was a mainstay of the Dietz line. This Blizzard is the improved version of the popular No. 1 Blizzard of Plate 6.15. The updated Blizzard has a large fill cap, LOC-NOB globe, more crimp construction,

Plate 7.15
Dietz Blizzard (1912), 1912-1945, 14.5 in.(36.8 cm) 6.75 in.(17.0 cm) $40-$120 USD

reinforced tubes, inside lift, and a standing bail with patented brass eyelet. The Blizzard is the largest and brightest hand lantern in the Dietz line.

Up to 1936 the Blizzard use the plain fill cap but as the new tooling became available, all Dietz lanterns began sporting the full text cap shown in Plate 7.15a.

"Made in America" had to be scrubbed from the dies when they were used to produce the Blizzard in the Hong Kong factory.

Variations:

Available in tin and metallic blue. The Blizzard can be found factory embossed with the owner's name. One of my favorites is * U.S.M.C.* From 1912 to 1962 Dietz built a lower cost twin to the Blizzard called the Dietz No. 2 Crescent.

Plate 7.15a
Dietz lanterns made in the U.S.A. say so on the fount and sometimes on the fill cap.

Plate 7.15b
The 1898, left, 1912, and 1939 Dietz Blizzards.

Plate 7.15c
The globe says "Blizzard" too.

Prisco Junior Wagon

Plate 7.16 **Description**:

The Junior Wagon is an all steel, cold blast lantern whose tin reflector has a 3 inch (7.6 cm) red glass in the center. This kerosene lantern has a steel, No. 1, rising cone burner, with a 0.625 inch (1.6 cm) wick, and uses a junior size globe. This lantern has a unique Prisco lift, round tubes, and a flat top fount. Construction is mostly machine crimp with some hand soldered parts. Finish appears to be black paint over a tin finish. This classic wagon lantern is a combination head light and tail light with a bracket for mounting to a wagon.

Plate 7.16
Prisco Junior Wagon, 1912-1920, 12.75 in.(32.3 cm) 5.5 in.(14.0 cm) $80-$90 USD

Markings: (on fount top)

PRISCO No 321(or 331)
COLD BLAST
MADE IN U.S.A.

Patent Information: (on the air chamber)

PATD JUN 13	05
· JANY 9	06
· APR 10	08
· JANY 13	12

(on the side bracket):

PAT. AUG. 1. 1911

Remarks:

The Prisco Junior Wagon lantern was made by the Pritchard Stamping Company of Rochester, New York. It is very much like the Dietz Junior Wagon of Plate 7.25. The globe used in this lantern is a junior Bull's Eye with the lens formed in the glass of the globe. An identifying feature of a wagon or buggy side lantern is the ruby glass mounted in the reflector. The side bracket has a side board spring and a receptacle for a wagon mounting bracket. The Prisco Wagon is not a dash lantern because it has no dash clip on the back (see the Buckeye Dash of Plate 7.2).

This Prisco has the advanced feature of a large, one piece, brass fount fill cap. This indicates this lantern was designed around 1912 when this became a popular feature. I was told this lantern came from an Alta-Dena Dairy wagon, Altadena, California. This example is complete and functional with globe and ruby lens intact. This lantern was stripped of paint and rust, repaired, primed, sealed and painted with black enamel. The dents that could not be easily removed, remain. Wagon lanterns have good collector appeal especially if they come from a small company like Prisco.

Variations:

The standard Prisco Junior Wagon is model 321. The model 331 is the same lantern with a 2 inch (5.1 cm) green glass lens on the left side. This is contrary to boat and airplane lighting where green is on the right.

Plate 7.16a
Here is the Prisco Model 321 on the left and the 331, with green side lens, on the right.

Prisco No 477 Cold Blast

Plate 7.17 **Description**:

The No. 477 is a steel, cold blast lantern with a large, one piece, brass fount cap, and a No. 2, brass burner with a 0.875 inch (2.2 cm) wick. The Prisco type of lift is unique. The original finish is unknown but other Prisco's (see Plate 7.16) have black paint over a tin coating. The correct globe is a No. 0 and the globe guards are similar to guards seen on Defiance lanterns.

Markings: (embossed on fount)

PRISCO No 477 COLD BLAST
MADE IN U.S.A.

Plate 7.17
Prisco No 477, ca., 1912, 15.25 in.(38.7 cm) 6.75 in.(17.0 cm) $60-$70 USD

Patent Information: (on side tube)

PATENTED

June	19,	05
Jan	9,	06
Feb	10,	08
May	4,	09
Jan	12,	12

Remarks:

Here is another lantern from Pritchard Stamping Company (PRISCO). It was found in the junk pile of an abandon cabin in the desert. There is not much chance of finding lanterns in dumps these days but sometimes they turn up in unexpected places.

In 1905, Albert R. Pritchard teamed with Henry Alvah Strong to found Pritchard - Strong Company in Rochester, New York. Charles T. Ham had been making lanterns in Rochester for 28 years when Pritchard started his stamping works. Strong had recently retired as the President of Eastman Kodak and was looking for a new venture. Strong did not stay long as the company name was changed to Pritchard Stamping in 1911. The lantern is marked with the PRISCO logo which is a contraction of Pritchard Stamping Company. No trace of the company is found in the Rochester directory after 1918.

The guard wires are not designed for a LOC-NOB globe so a generic No. 0 globe is appropriate. Perhaps Prisco purchased globes from nearby Corning Glass Company, the inventors of Pyrex® heat resistant glass.

Variations:

This same lantern, with a brass fount, is marked:
RAYO No. 82 COLD BLAST MADE IN U. S. A.

Rayo marked lanterns were made by Embury, Defiance, Dietz, as well as Pritchard Stamping.

PRISCO

PRITCHARD
STAMPING
COMPANY

Plate 7.17a
Pritchard Stamping Company was formerly Pritchard - Strong Company which also abbreviates as PRISCO.

Dietz No 2 D-Lite

Plate 7.18 **Description**:

This all steel, cold blast lantern has a large, two piece fount cap embossed with the Dietz logo. It has a "short" globe, inside lift, all steel, No. 2, domed burner, with a 0.875 inch (2.2 cm) wick and bent wire wick adjust. The air tubes are machine embossed for strength. Machine crimp construction with some solder. Advertised as burning 20 hours while giving 10 candlepower, the D-Lite was made in New York City and Syracuse, New York.

Markings: (embossed on fount)

DIETZ N.Y. -U.S.A.
H No 2 D-LITE H

(embossed on crown and fill cap)

DIETZ (in oval)

(embossed on globe)

DIETZ N.Y. U.S.A.
NO. 2 D-LITE

Patent Information: PATENTED

MAY - ?? - 20
DEC - 27 - 32
S - 4 - 36

(the first D-Lite is marked)

MCH - 3 - 06
NOV - 19 - 07
MAR - 8 - 10
JUNE - 21 - 10
MAY - 7 - 12
JULY - 12 - 1

Plate 7.18
Dietz D-Lite (1913), 1913-1944, 13.25 in.(33.7 cm) 7.75 in.(19.7 cm) $60-$120 USD

Remarks:

The D-Lite lantern first appeared in 1912 as C. T. Ham's bold "Nu-Style" lantern designed by Warren McArthur Jr. The lantern has many new features but the boldest must be the "short" globe. One advantage of the short globe is a hand will fit inside for easier to cleaning. When Dietz bought the dies from the closed Ham company in 1915, they renamed this lantern D-Lite.

There are two sizes of founts built between 1919 and 1944. The smaller was 6.75 inches (17.2 cm) diameter and the larger was 7.75 inches (19.7 cm) in diameter. Both versions are 13.25 inches (33.7 cm) tall. The "special" D-Lite has a larger fount that doubles the burning time to 40 hours.

The D-Lite was redesigned in 1939 to produce the deco lantern of Plate 8.10. The 1919 D-Lite tooling was used again in Hong Kong to make the D-Lite No. 90 of Plate 8.18.

Variations:

The Ham Nu-Style was designed with the most advanced features and was the most expensive cold blast lantern in the Dietz line at the time. It had the new No. 282 rising cone burner and the new short globe. They were available in bright tin or with a brass fount and crown. The higher cost may explain why the first D-Lite is difficult to find and why they were redesigned in 1919. The revised D-Lite is nearly identical to the existing No. 2 Wizard, so the Wizard name was dropped. D-Lite globes were available in clear, ruby, blue, and green.

Plate 7.18a
From 1913 until 1918, the Nu-Style lift was on the chimney and a rising cone burner was used.

Dietz Monarch (1913)

Plate 7.19 **Description**:

The 1913 Dietz Monarch is an all steel, hot blast lantern, with a No. 411, wing-lock, steel burner (0.625 inch wick). It has a plain (no logo), two piece, fount cap with the top half steel and the bottom half brass. The Monarch has a rear lift, high dome fount top, and machine crimp construction with some soldered parts. The air tubes are embossed for strength when pressed in the stamping machine. It has a cross wire globe guard and uses a clear or ruby FITZALL or LOC-NOB globe. This very successful general purpose lantern, made in New York, is said to produce 4 candle power for 18 hours.

Markings: (on fount)

DIETZ NEW YORK U.S.A.
MONARCH
(on crown)
DIETZ (in oval)
(on globe obverse)
DIETZ
FITZALL
N.Y. U.S.A.
(on globe reverse)
H3
LOC-NOB
PAT'D 3-10-14

Patent Information: (on center tube)

PATENTED
JULY 11 - 9?
JULY 26 - 04
NOV. 19 - 07
JUNE 21 - 10
SEPT. 16 - 13
S - 10 - 29

Remarks:

The Dictz Monarch was updated in 1913, just in time for World War One. This Monarch was faster and easier to build because it uses more machine operations than the previous Monarch model (see Plate 6.18). This lantern continued to be made until about 1950.

The correct globe is the LOC-NOB, FITZALL globe that fits snugly into the globe guard wires. The air tubes are machine embossed for strength. The correct finish for this lantern is bright tin plate.

This Monarch was built after 1918 as the globe guard wires were shaped differently before 1918.

The 1913 Monarch was the second tooling of what would become the most successful line of hot blast lanterns ever produced...the Dietz Monarch. Compare this lantern to the 1936 streamline version in Plate 8.3, and the 1986 reissue in Plate 8.19.

Variations:

Most were tin plated. The globe guard wires had a different shape on Monarchs made from 1914 to 1918.

Put a No. 2 burner in a Monarch and you get a Dietz No. 2 Royal. The No. 2 Royal is 13.5 inches (34.3 cm) overall, and advertised as giving 5 candlepower.

The Dietz Little Star is an 11 inch (27.9 cm) tall lantern with square tubes. The Little Star uses the same burner and wick but the globe is only 5.5 inches (19.0 cm) tall. The Little Star U.S. globe is used only on the Dietz U.S. and Little Star lanterns.

Plate 7.19a

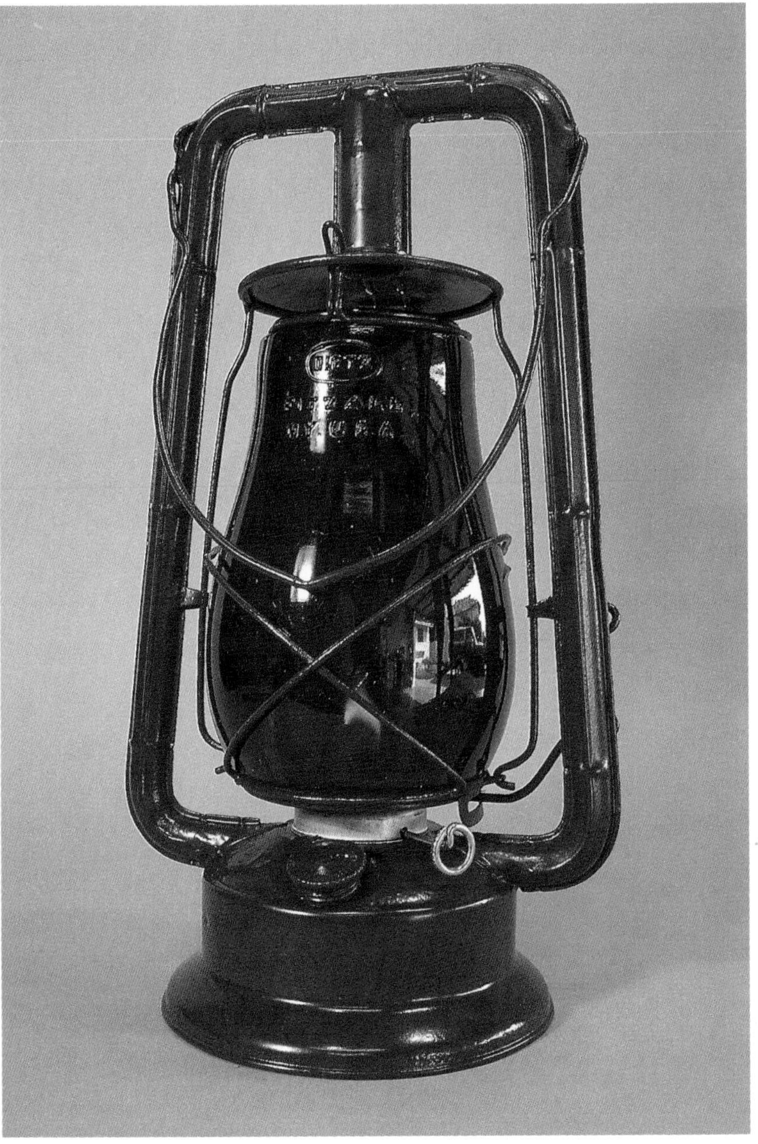

Plate 7.19
Dietz Monarch (1913), 1913-1950, 13.5 in.(34.3 cm) 6.0 in.(15.2 cm) $40-$50 USD

Plate 7.19b
The FITZALL globe fits just about everything.

Adlake Reliable

Plate 7.20 **Description**:

The Reliable is a dead flame brakemen's lantern with a brass burner and a 0.625 inch (1.6 cm), cotton wick. The balance of the lantern is crimped, electrically welded, and riveted steel. This lantern is typical of railroad style lanterns from 1880 to 1960. It has an insert pot and burner that drops in from the top. Access to the lantern wick is by opening the top and lifting the globe. Original finish appears to be tin over the steel. This lantern uses the number 39 globe.

Markings: (on burner)
ADLAKE
PAT.
APRIL 27, 1909
(on crown)
ADLAKE RELIABLE
THE ADAMS AND WESTLAKE CO.
NEW YORK CHICAGO PHILA.
UNION PACIFIC

Patent Information: (around top on model with two part bottom)
PATENTED MAY 5, 1908
- 2-PATS. JAN 26, 1909-NOV. 28, 1911 -2- PATS. JULY 2, 1912 APR. 1, 1913
(variation on model with one part bottom)
PAT. 5.5.1908-2- PATS 1.26.09-11.28.1911-7.2.1912-4.1.1913-5.9.1922-CAR.1921-1923-PATS. PENDING

Remarks:

The Adlake Reliable was extensively used by railroads, utilities, the military and the public. Just because a lantern looks like a railroad lantern does not prove it was ever owned by a railroad. The exception is any lantern factory embossed with the owner's name. This example is marked for the Union Pacific railroad of Cheyenne, WY.

This lantern is common, plain and unremarkable but typical of the hundreds of thousands of railroad style lanterns produced in U. S. factories.

This lantern was made to use the No. 39 globe. The original globe color would determine the lantern's intended use. Clear globes were used for general lighting, red for danger or signaling. The use of the colors blue, green, and amber would be determined by each railroad's needs. Even a Bull's eye globe was available. Globes are often found marked with the railroad's name.

A note of caution. Etching of globes with popular railroad companies names or logos is easy to do by unscrupulous parties. If the logo or name is embossed in the tin or glass then it has added collector value but, if it is etched in the globe, take it with a grain of salt.

The No. 39 globe is sometimes referred to as a tall globe because they are taller than the kero globe of Plate 8.1. The short kero globe is still being made and this has kept the price of short globe lanterns lower.

Variations:

There are many lanterns similar to the Reliable and the details of the Reliable changed over the years. A good source of detailed lantern information can be found in The Illustrated Encyclopedia of Railroad Lighting by Richard C. Barrett.

From 1865 to 1905, railroad style lanterns that burned sperm oil used a barrel shaped, 6 inch (15.24 cm) globe. This lantern can take an early 6 inch or the 5.38 inch (13.6 cm) tall No. 39 globe.

Plate 7.20
Adlake Reliable, ca., 1913, 10.0 in.(25.4 cm) 7.0 in.(17.8 cm) $100-$110 USD

Plate 7.20a
The original owner of this Adlake Reliable was the Union Pacific Railroad.

Plate 7.21
Adlake RR Semaphore, ca., 1913, 15.5 in.(39.0 cm) 5.25 in.(13.0 cm) $90-$100 USD

Adlake Semaphore Lantern

Plate 7.21 **Description**:

This dead flame lantern has a nickel plated, brass burner with a 0.25 inch (0.64 cm) round, felt wick. The balance of the lantern is crimped, soldered and riveted steel. This lantern is unusual in that it has a single, clear, 5.25 inch (13 cm) lens and no provisions for holding or using colored filters. The fount is small and slips into the bottom on rails. The only access to the lantern's fount is by sliding the entire front up. The back has a 0.75 inch (1.9 cm), round flame viewing window. Original finish appears to be black paint over the steel. It produces a very directional, yellow light. From the collection of the Belmont Shores Model Railroad Club, Los Angeles, California.

Markings: (on wick adjust)
· ADLAKE · CHICAGO
(inside on paper label)
THE ADAMS AND
WESTLAKE CO
CHICAGO ILL.

Patent Information: (inside on brass disk)
PATD IN US JAN 24 1899 JUN 12 1900
JAN 13 1903 AUG 4 1903 OCT 27 1903
MAR 29 1904 AUG 13 1907 DEC 24 1907
MAR 16 1909 APR 27 1909 JUL 16 1912
AUG 6 1912 SEP 24 1912 JAN 21 1913
DEC 2 1913
OTHER PATS PEND
THE ADAMS & WESTLAKE CO
CHICAGO NEW YORK PHILADELPHIA

Remarks:

This Adlake railroad semaphore lantern uses a single lens to produce a focused light. The clear lens appears to be original. The mounting brackets is designed for a permanent location. Compare to the Adlake Switch Lantern of Plate 79. This small, permanent mount, single lens lantern was designed for use as a semaphore light. It is mounted on a mast, behind the moving target of the semaphore which holds the colored lenses.

The patent dates suggest it was built in the late teens or early 1920s.

Plate 7.21a
Railroad semaphore of the type that would be lit by this lantern.

Plate 7.21b

Plate 7.21c
This big lantern has a surprisingly small wick and fount.

Warren STA-LIT

Plate 7.22 **Description**:

The Sta-Lit No. 2 is all steel, cold blast lantern with a large one piece fount cap. The No. 2 uses a short globe and has a brass, No. 2, domed burner, with a 0.875 inch (2.2 cm) wick. The bale is attached to the air tubes in a unique way and the Sta-Lit has an inside lift. The large fount and paint color (red) give the impression the Sta-Lit was intended for contractor use. Machine crimp and hand solder construction is used in the Sta-Lit.

Markings: (embossed on fount)

STA-LIT No. 2 C. B.
THE WARREN STAMPING CO.
WARREN O. U.S.A.

Patent Information: (on crown)

PAT
JAN. 10, 05
JAN. 17, 11

Remarks:

In 1908, Brothers Frank A. Iddings and William T. Iddings started the Iddings Company in Warren Ohio. The name was changed to Warren Stamping Company between 1912 and 1915. Lanterns were only a small part of their business. Warren Stamping also produced household products like flour sifters and steam cookers.

If the Sta-Lit is a contractor's lantern it has an unusually large burner. A large burner uses fuel faster and contractors usually want more economy.

The Sta-Lit was made by an independent metal stamping company and it has a few unique features. One of the two patents on the lantern is likely for the unusual way the bale is attached to the tubes. It is wrapped around the tube so there is no need to make any holes. The advertising says: "Bail stays in any position" and it works quite well.

The original finish for this lantern appears to be red paint and the fill cap is identical to the PRISCO caps of Plates 7.16 and 7.17.

Compare this lantern to the other lanterns that use the short globe; the Dietz D-Lite of Plate 7.18, the Embury Air Pilot of Plate 8.4, and the BUHL Little Giant in Plate 7.23.

Plate 7.22
Warren STA-LIT, ca., 1914, 14.0 in.(35.6 cm)
7.75 in.(19.7 cm) $60-$70 USD

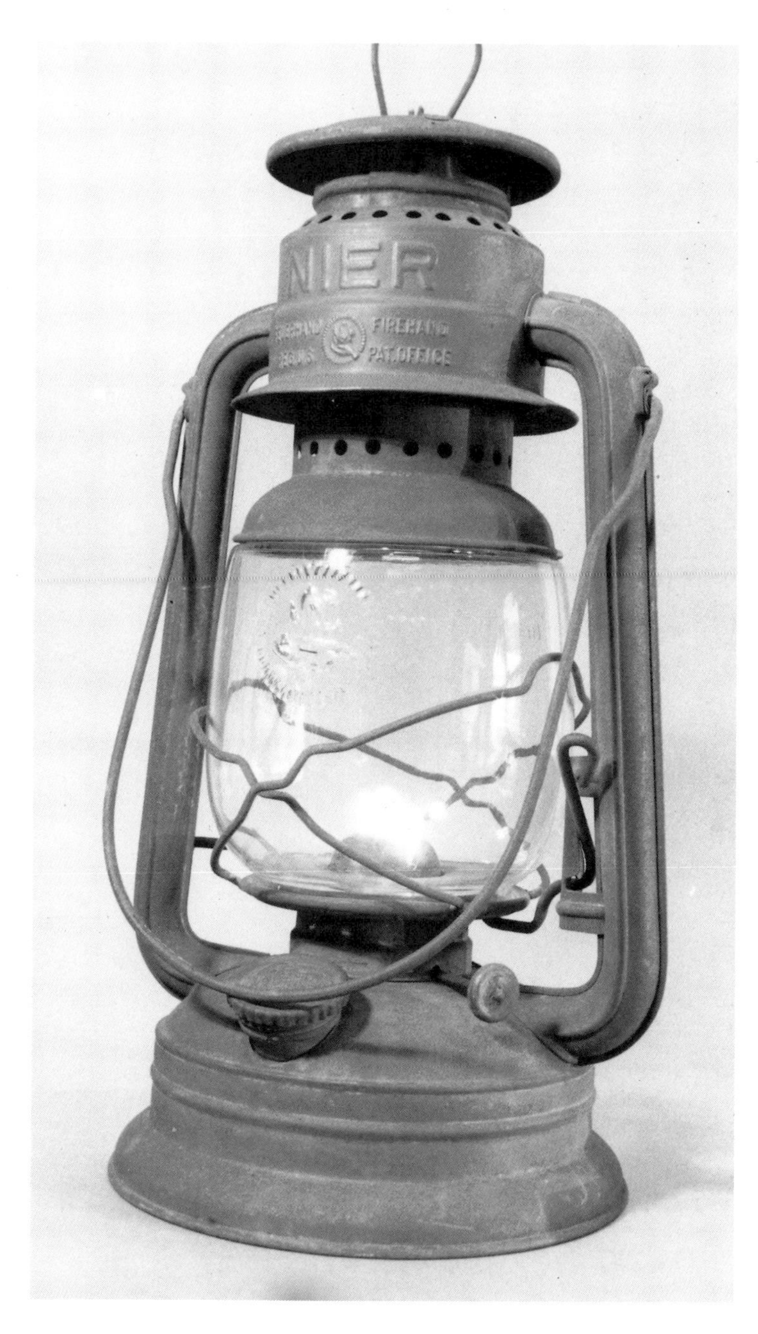

Plate 7.22a
The short globe Warren McArthur Jr. designed in 1912 became quite popular and is even seen on foreign lanterns like this pre-war Feuerhand from Germany.

Buhl Little Giant

Plate 7.23 **Description**:

The Buhl No. 875 is a steel, cold blast lantern with a brass fount. It has a large, two piece, brass fount cap, a brass, No. 2 burner, a 0.875 inch (2.2 cm) wick, an inside lift, a slightly domed fount with machine and hand solder construction. The correct globe is a "short" globe. The large, brass fount provided corrosion resistance and the balance of the lantern is bright tin. Made in Detroit, U.S.A.

Markings: (embossed on bell)
BUHL No 875
(embossed on fount)
LITTLE GIANT
DETROIT U. S. A.
Patent Information: (none)
Remarks:

Here is an interesting find for two reasons. First this lantern has a solid brass fount. Second, it was made by a smaller company in the motor city.

The Buhl Stamping Company was started in 1888 by the directors of Buhl, Sons, & Company, a wholesale hardware supplier. The President was a Theodore D. Buhl. The Buhl company's original business was the production of several sizes of milk cans but within five years the lantern capacity was 5000 per month. C. T. Ham is known to have sold parts to the Buhl factory. Declining sales and the Great Depression ended lantern production but Buhl continued to produce stamped steel products until 1956.

The Little Giant has a brass fount. Using brass for the fount is not unusual because rusted out founts was a common complaint. Rust occurs in any lantern left setting on the ground where moisture can collect on the bottom. Other solution for the rust problem can be seen in the Embury Camlox of Plate 7.13 (cop-

Plate 7.23
Buhl Little Giant, 1914-1925, 13.0 in.(33.0 cm) 6.75 in.(17.0 cm) $100-$120 USD

per) and the Dietz Crystal in Plate 6.17 (glass). Another lantern available with a brass fount is the Dietz D-Lite of Plate 7.18. The disadvantage was, of course, that special founts cost more to buy so they were made on smaller quantities. Smaller quantities mean unusual founts are more difficult for collectors to find.

This lantern is called "Little Giant" because it is shorter than a lantern using a No. 0 globe, but it still has the large, No. 2 burner. It is about the same size as the Dietz Wizard. Do not confuse this lantern with the even shorter Dietz Little Giant (Plate 7.31) that used an even smaller globe.

Compare this lantern to the other lanterns that use the short globe; the Embury Air Pilot of Plate 8.4, the Dietz D-Lite of Plate 7.18, and the STA-LIT in Plate 7.22.

Variations:

Buhl produced a full line of hot and cold blast, dash, wagon, side, and inspector lanterns. Each model has a name like Majestic, Eclipse, Conquest, and Princess.

BUHL
LANTERN COMPANY
Detroit, Michigan, U.S.A.

Plate 7.23a

Defiance Perfect

Plate 7.24 **Description**:

The Defiance Perfect is an all steel, hot blast lantern of conventional design. The burner is a steel, No. 1 with a 0.625 inch (1.6 cm) wick. This lantern has an inside lift and a large, two piece, steel fount cap. There is no sign of paint so the original finish was probably tin. The Perfect was made in Rochester, N.Y. using machine crimp processes with hand soldered parts.

Markings: (embossed on crown)

MADE IN U.S.A.

(embossed on fount)

DEFIANCE LANTERN & STPG. CO.
ROCHESTER N.Y. U.S.A.
No 0 PERFECT

(embossed on the globe)

D. L. & S. CO.
PERFECT

Patent Information: (none)

Remarks:

Various forms of the Perfect were produced by Defiance from 1900 to 1930. The Perfect was a simple and inexpensive lantern that sold well.

William C. Embury gained his lantern experience working for the Kemp Manufacturing Company in Toronto, Ontario, Canada. In 1900 he left Kemp to start the Defiance Lantern and Stamping Company in Rochester, N. Y. Embury was a very successful salesman for the Defiance company but disagreements over his partners hiring practices forced his departure in 1908. Embury started the Embury Manufacturing Company in the same town that year. The Embury company soon surpassed Defiance and in 1930, purchased the assets and equipment of Defiance Lantern and Stamping. Once again proving, "Success is the best revenge."

The 1900 to 1915 Perfect was nearly identical to the 1905 Dietz Monarch of Plate 20. The Perfect is similar to the Defiance No. 0 Regular.

Variations:

This and other Defiance lanterns were available with a ruby or clear globe.

Plate 7.24
Defiance Perfect, 1914-1930, 13.25 in.(33.7 cm) 6.25 in.(15.9 cm) $60-$70 USD

Plate 7.24a
Defiance Perfect looking less than perfect before its chemical bath and two coats of paint.

DEFIANCE LANTERN
& STAMPING CORP.
ROCHESTER NEW YORK U.S.A.

Plate 7.24b

Dietz Junior

Plate 7.25 **Description**:

The Dietz Junior is an all steel, cold blast lantern. It has a small, one piece fount cap and brass, No. 1, domed burner with a 0.625 inch (1.6 cm) wick. The Junior has an inside lift and one of the earliest examples of a dome fount. The fount seam is double crimp construction. The globe is clear and embossed with a very faint DIETZ N.Y. U.S.A. JUNIOR.

Plate 7.25
Dietz Junior, 1914-1956, 12.25 in.(31.0 cm) 5.25 in.(13.3 cm) $20-$100 USD

Markings: (embossed on fount)
DIETZ JUNIOR N.Y. U.S.A.
(Sanskrit Text)
(embossed on globe)
DIETZ N.Y. U.S.A.
JUNIOR
Patent Information: (on side tube)
PATENTED

FEB	-1	-98
JULY	-11	-99
AUG	-7	-00
JULY	-26	-04
NOV	-19	-07
MAR	-8	-10
JUNE	-21	-10
MAY	-7	-12
SEP	-16	-13
SEP	-9	-13

Remarks:

The Dietz Junior began in 1898 and continued to the present day. All junior lanterns use the junior size globe. Dietz claimed there were more Dietz Juniors in use around the world than any other brand.

This example has Indian Sanskrit text and was made after 1913. The Sanskrit text indicates this lantern was intended for export to India. I was told this one came from Kansas. Why it did not go to India is not known.

The patent dates suggest it is 15 years older than the Junior Wagon of Plate 7.25. The only difference in construction is the later lantern has a larger fount cap.

Variations:

Because the Dietz Junior was a more gentile lantern it was available in bright tin, brass, and nickel plated brass. Variations include outside lift before 1913. The junior was updated in 1939, along with the rest of the Dietz line. The 1939 model is painted metallic blue. The finish in later years included brass, nickel, tin, red, and black. This lantern is still available today as a Chinese import.

Plate 7.25a
The Indian export models can be identified by the Sanskrit text on the fount.

SANSKRIT - अमीरिका
आ ल्लिडाट्ज्ज्निमाट
TRANSLATION -
AMERICA
REAL DIETZ JUNIOR

Plate 7.25b
Sanskrit message on the Junior fount says this lantern is Made in America.

Dietz Junior Wagon

Plate 7.26 **Description:**

This cold blast lantern has a steel, No. 1, dome burner with a 0.625 inch (1.6 cm) wick. The Junior uses an inside lift, and has a dome fount. Construction is mostly machine crimp with some hand solder on the reflector. As with all wagon lanterns, this Junior has a 2 inch (5.1 cm) ruby rear lens in the reflector and a 2.125 inch (5.4 cm) bull's eye lens up front. All wagon lanterns have a provision for mounting to a special bracket on the left or right side of a buggy, surrey or wagon.

Markings: (embossed on crown)
DIETZ (logo)
(embossed on reflector)
DIETZ (logo)
JUNIOR WAGON
LANTERN N. Y. U. S. A *
(embossed around burner dome)
DIETZ JUNIOR *
N.Y. U.S.A. *
(embossed on fount)
DIETZ JUNIOR N.Y. U.S.A.
(and Sanskrit text)
(embossed on globe)
DIETZ (logo)
JUNIOR COLD BLAST
MADE IN U. S. A
Patent Information: (on side tube)
PATENTED

FEB	-1	-98
JULY	-11	-99
AUG	-7	-00
JULY	-26	-04
NOV	-19	-07
MAR	-8	-10
JUNE	-21	-10
MAY	-7	-12
SEP	-16	-13
SEP	-9	-13
S	-8	-28

Plate 7.26
Dietz Junior Wagon, 1914-1945, 12.0 in.(30.5 cm) 5.25 in.(13.3 cm) $60-$100 USD

Remarks:

The Dietz Junior Wagon uses a magnifying lens attached to the globe plate instead of a bull's eye globe. The correct globe for this lantern is the Dietz junior size globe. Another identifying feature is a ruby glass mounted in the reflector. Unlike the front lens, the rear lens does not magnify.

The patent date suggests it is 15 years newer than the Junior of Plate 7.24. The only improvement evident is a large, two piece, fount cap which, if you ever try to fill one, is a big improvement.

This example has a clip for left side mounting on a wagon. It has no dash clip on the back (see the Buckeye Dash of Plate 7.2). This is the essential difference between a dash lantern and a wagon lantern. A wagon lantern attaches at the side so the red light can be seen from the rear. A dash lantern clips to the dash and therefore needs no rear lens.

The Sanskrit text indicates this lantern was intended for export to India. The Sanskrit says this lantern is made by Dietz in America.

Variations:

The original finish was japanned black only. A similar lantern was available without the reflector (see Plate 7.25). The Dietz Junior Dash made from 1898 to 1920 has an outside lift.

Plate 7.26a
Rear view of the Junior Wagon shows its red glass.

Plate 7.26b
Tag found on all Dietz lanterns of the period.

Dietz Eureka Buggy

Plate 7.27 **Description:**

The Dietz Eureka is a dead flame, steel buggy lantern with some brass parts and a nickel plated lens bezel. The No. 561 burner is brass with a 0.625 inch (1.6 cm) wick. Construction is mostly machine crimp with some hand soldering. The buggy lantern has a 3 inch (7.6 cm) front lens, a 2 inch (5.1 cm) ruby rear glass, and a 1.5 inch (3.8 cm) green side lens. The Eureka has a small bail and a mounting bracket with a locking screw. The fount is removable with a twist and the wick adjust is external. The fount is stuffed with cotton for safety and one fill burns for 10 hours.

Markings: (embossed on crown)

* DIETZ EUREKA *
MADE IN U.S.A.
PATENTED *
JULY-26-04 *
MAY-19-08 *
FEB-10-14 *

Remarks:

The Dietz Eureka is advertised as a buggy or auto side lantern but it came too late to be popular. By 1914 buggies were quickly becoming a thing of the past and all new autos (with the notable exception of the Model T Ford) had electric lights. The market for another kerosene side light was limited. The Eureka lacked the style of the fancy auto side lanterns or the sophistication of adding electric lights to the old family auto. The Eureka is smaller than a tail light (see Plate 7.6) and larger than a bicycle lantern. Dietz tried to sell the Eureka as a parking light for autos by advertising, "Saves battery current."

This lantern uses a clear magnifying lens-in-door at the front, a green magnifying lens on the left side, and a ruby glass on the rear. The rear lens does not magnify. The right side has a bracket and locking screw. The Eureka has a small bail for use as a hand lantern. The front bezel is nickel plated brass. Chrome was not in common use until the late 1920s. The balance of the finish is black enamel paint. The front lens of this lantern is crizzled but could be replaced. The steel parts could be cleaned and repainted and the nickel part replated.

A dozen of these buggy lanterns wholesaled for $38.00 in 1926.

Variations:

Buggy and auto tail lights are similar in appearance to the Eureka, but larger. Bicycle headlights were slightly smaller but worked on the same principle. The Eureka comes in left and right side mount.

Plate 7.27
Dietz Eureka, 1914-1930, 7.5 in.(19.0 cm)
5.0 in.(12.7 cm) $100-$150 USD

Plate 7.27a
Red glass on the rear of the Dietz Eureka.

Embury No. 160 Supreme

Plate 7.28 **Description**:

The Embury Supreme No. 160 is an all steel, cold blast, lantern for kerosene. It has a two part, large, steel fount cap, and a No. 2, steel, rising cone burner with a 0.875 inch (2.2 cm) wick. This Supreme has an inside lift and was originally plated with tin. The Supreme uses a short globe and is marked for a distributor named Ward. Built in Warsaw, N.Y. using machine crimp and hand soldered parts. *From the collection of Gilbert Belcher.*

Plate 7.28
Embury No. 160 Supreme, 1916-1952, 14.5 in.(36.8 cm) 7.25 in.(18.4 cm) $10-$20 USD

Plate 7.27b
The lens is hinged to allow cleaning.

Markings: (embossed on fount)

WARD'S STANDARD
No. 2

(alternate embossed on fount)

EMBURY SUPREME
NO. 160

Patent Information: (none)

Remarks:

The Embury Manufacturing Company was started in 1908 by William C. Embury, and moved to Warsaw in 1911. Embury's sons closed the doors on the last day of 1952.

This is another example of a lantern marked for a distributor and not the manufacturer. Embury sold much of their annual production custom marked for jobbers, distributors, or municipal utilities. Embury lanterns can be found with a wide variety of markings. This makes lantern collecting just a little more challenging as the company name, city, and state embossed in the steel, may be a red herring. This lantern is an Embury Supreme No. 160 but it is factory embossed "WARD'S STANDARD." The "No. 2" refers to the burner size.

Variations:

Another name found on Supreme is a hardware supplier in Louisville KY:

BELKNAP HWD. & MFG. CO.
BLUE GRASS

Unfortunately, Belknap also called the Embury No. 2 "Bluegrass," which is just plain confusing.

This lantern, and all the other Embury lanterns, were available with ruby or clear globes. Compare to the smaller Little Supreme in Plate 7.30.

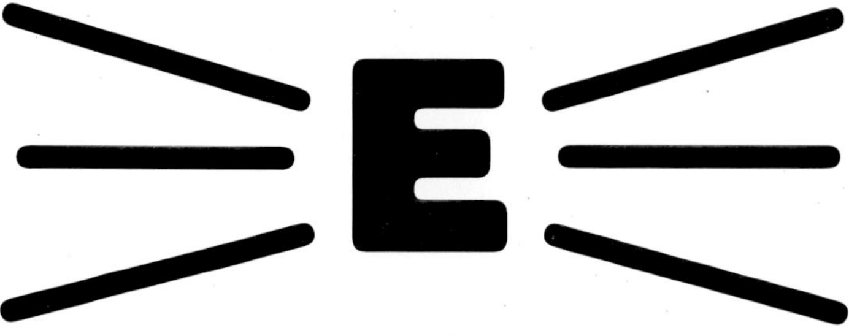

Plate 7.28a
The "E" logo found on Embury globes.

Plate 7.28b
Back side of the No. 160.

Defiance No. 200

Plate 7.29 **Description**:

The Defiance No. 200 is a large, all steel, cold blast lantern with a larger than usual fount. The No. 200 has a steel, No. 2, domed, burner with a 0.875 inch (2.2 cm) wick. This lantern has an inside lift and a large, two piece, steel fount cap. There is no sign of paint so the original finish was probably tinned steel. The globe guard style does not suggest the use of a LOC-NOB globe.

Markings: (embossed on fount)

DEFIANCE LANTERN & STPG. CO.
ROCHESTER, N.Y.
NO. 200 DEFIANCE

(embossed on globe)

D. L. & S. Co.

Patent Information: (none)

Remarks:

The No. 200 is a big lantern. At 15 and 0.75 inches (40.0 cm) it is the largest hand lantern in this book.

The Defiance Lantern and Stamping Company was organized by William C. Embury and Canadian partners in 1900. The name came from their resolve to defy the might of the R. E. Dietz and the other large lantern producers. Disagreements with the Canadian partners over nepotism lead to the departure of Bill Embury in 1908. Embury went on to start the Embury Manufacturing Company (see Plate 7.28).

The Defiance Lantern and Stamping Company had globes produced that are marked D. L. & S. Co. These globes are hard to find so a generic No. 0 globe could be used until the correct globe is located.

This specimen is similar to the Prisco No 477 in Plate 7.17. Compare the largest to the smallest lantern, the Kwang Hwa, in Plate 9.8.

Variations:

Defiance also produced a Defiance Inspector's lantern similar to the Dietz ACME of Plate 7.7, as well as number 39 Signal Lantern like Plate 7.10. Also see the very popular Defiance Perfect hot blast lantern in Plate 7.24.

Plate 7.29
Defiance No. 200, ca., 1919, 15.75 in.(40.0 cm)
7.5 in.(19.0 cm) $20-$30 USD

DEFIANCE LANTERN
& STAMPING CORP.
ROCHESTER NEW YORK U.S.A.

Plate 7.29a

Embury Supreme No. 210

Plate 7.30 **Description**:

The Embury Supreme is an all steel, hot blast, kerosene lantern. It has a two piece, large, steel fount cap, and a No. 1, steel, domed burner with a 0.625 inch (1.6 cm) wick. This Supreme has an inside lift and is finished in tin. The globe is a clear, Embury No. 0.

Markings: (on fount)

NO. 210 SUPREME
EMBURY MFG. Co.
SUPREME
WARSAW, N.Y. U.S.A.

Patent Information: (none)

Remarks:

The Embury Manufacturing Company of Rochester, New York was incorporated November 27, 1908 by William C. Embury. The company designed and produced a variety of tubular and railroad style lanterns. The factory moved to Warsaw, N.Y. in 1911, where it remained until R. E. Dietz Company purchased it in 1953.

There is another lantern in the Embury line named Supreme. Do not confuse this Supreme with the Little Supreme of Plate 7.30.

The design of this lantern places it after W. W. 1 and before the Deco restyling that swept the lantern industry in the late 1930s. The Supreme has a clean and simple design that would have been inexpensive to purchase. This lantern is ideal for camping, hunting, and fishing as well as for use around the farm.

This example was in excellent condition when found. It had no signs of original paint so the finish was most likely tin plate.

Despite the similar appearance, Embury fill caps are not interchangeable with Dietz caps.

It was very common for the Embury Company to build a special run for jobbers, distributors, utilities and municipalities. These lanterns can be found with a wide variety of markings. For just a small sample of these jobbers refer to the Embury No. 2 Air Pilot in Plate 8.4.

Variations:

This lantern, and all the other Embury lanterns, were available with ruby or clear globes. Paint colors could be special ordered. Compare this lantern with the completely different Little Supreme of Plate 7.30.

Plate 7.30
Embury No. 210 Supreme, 1919-1938, 13.25 in.(33.7 cm) 6.0 in.(15.2 cm) $20-$30 USD

EMBURY MFG. CO.
(SUPREME)
WARSAW N.Y. U.S.A.

Plate 7.30a

Embury Little Supreme

Plate 7.31 **Description**:

This all steel, cold blast lantern uses a little wizard size globe. It has a steel, No. 0, dome-less burner, 0.625 inch (1.6 cm) wick, Embury loop wire wick adjust, round tubes, and an inside lift. Construction is all machine crimp. Finish is the original Embury green metallic paint. This contractor's lantern has a ruby globe but clear globes were also available.

Plate 7.31
Embury Little Supreme, 1920-1952, 12.75 in.(32.3 cm) 7.25 in.(18.4 cm) $20-$30 USD

Markings: (on fount top)
No. 350 LITTLE SUPREME
EMBURY MFG CO.
SUPREME
WARSAW, N. Y. USA
Patent Information: (none)
Remarks:

The Embury Little Supreme was manufactured from 1945 to 1962. The Embury Manufacturing Company was incorporated in 1908 by William C. Embury and by 1923 it had 175 employees and sales of 900,000 units. In December 1930, Embury purchased the Defiance Lantern and Stamping Company. Bill Embury Sr. passed away on June 11, 1943 and the Embury company was dissolved by Phil, Fred, and Bill Jr. on the last day of 1952. The R. E. Dietz Company purchased the tools and equipment in 1953 and continued to sell some of the more popular lanterns.

The Little Supreme, with a ruby globe, was used by contractors as safety lights around construction projects. The practice slowed when, in 1955, the Federal Government passed a law forbidding flames to be used on government contracts. Lanterns and kerosene "torches" continued to be used around private construction until vandalism and theft made it impractical in the mid 1960s.

Many of these lanterns were purchased in large numbers by municipalities. In quantity, lanterns could be special ordered with the city or utility name embossed in metal. It is not unusual to see Embury lanterns factory embossed with STREET DEPT., D W & P, or PROPERTY OF (city name). Contractor lanterns have poor collector appeal. This lantern is common.

Variations:

The Little Supreme was restyled in the mid 1930s and renamed Little Air Pilot (Plate 8.5). Note both share the same globe, burner, fill cap, bail, crown, lift and other parts.

Plate 7.31a

Dietz Little Giant

Plate 7.32 **Description**:

The Dietz Little Giant (not to be confused with the Buhl Little Giant) is an all steel, cold blast lantern that claims a 70 hour fount capacity. It usually has a ruby wizard size globe. The Little Giant uses a steel, No. 1, domed burner, with a 0.625 inch (1.6 cm) wick and bent wire wick adjust. The large, two piece fount cap is embossed with the full Dietz text after 1936. The air tubes are stamp embossed for strength. The construction is machine crimp with and a touch of solder where the tubes meet the fount. The finish is usually red paint for contractor's use.

Markings: (embossed on crown)
DIETZ (logo)
(embossed on fount)

Plate 7.32
Dietz Little Giant, 1920-1962, 11.75 in.(29.8 cm) 7.75 in.(19.7 cm) $10-$20 USD

DIETZ N. Y. - U.S.A.
LITTLE GIANT
70 HOUR FOUNT CAPACITY
(on bell)
PROPERTY OF L. A. COUNTY
(on globe)
DIETZ (logo)
SYRACUSE, N.Y. U. S. A.
LITTLE WIZARD
LOC-NOB
REG'D. U. S. PAT. OFF.
(full text cap)
Patent Information: (on side tube)
PATENTED
MAY ?? 20
DEC 27 32
MADE IN U.S.A.
?? - 8 - 39
Remarks:

The Dietz Little Giant is a 1920 Little Wizard with a larger fount. The Little Giant, Little Star, and Little Wizard are the small lanterns of the Dietz line. The Little Star is a small Monarch, the Little Wizard is a small Wizard, and the Little Giant is a small Wizard with a larger fount.

To further confuse things, the Little Giant was renamed Little Wizard No. 1 and imported from Hong Kong (see Plate 8.16).

This lantern is commonly found and easy to identify by its short stature and broad fount. The large fount allowed it to burn for up to 70 hours without maintenance so it was very popular with municipalities and utilities.

Since this lantern was primarily used as a safety light, it is usually found painted red with a ruby, little wizard globe. The little wizard globe was a popular size for contractor's lanterns starting in the 1920s and continuing through the 30s, 40s and 50s.

As a cost cutting measure the little wizard globe was changed to a smooth, clear glass coated with a red, high temperature paint. There are clear patches left clear for viewing the flame. This paint blocked the light and the red paint soon faded to pink in the sun.

Variations:

The Little Giant was produced in large numbers, often with the name of the owner embossed on the bell. These lanterns are usually found with name of a city street department, Gas Company, phone company, or DWP, stamped in the bell.

The Little Giant was so successful that Dietz reissued it virtually unchanged as the Dietz Little Wizard No. 1 (Plate 8.16). These lanterns were updated in 1939. Compare to the No. 100 of Plate 8.7 and the Little Wizard in Plate 8.16.

The standard color was red but a Little Giant could be special ordered in gray, blue, or just about any color.

Wheeling Leader

Plate 7.33 **Description**:

The Paull's Leader is a large, all steel, cold blast lantern with a 32 ounce fount. The burner is a steel, No. 2, with a 0.875 inch (2.2 cm) wick. This lantern has an inside lift and a large, plastic fount cap. There is no sign of paint so the original finish was probably tin.

Markings: (embossed on fount)

PAULL'S LEADER *
LARGE No 2 FOUNT *

Patent Information: (none)

Remarks:

The Nail City Lantern Company of Wheeling, West Virginia was started in 1877 by Archibald Woods Paull, Sr. Nail City produced a variety of dead flame railroad style lanterns. By 1895, production had reached 184,500 lanterns. In 1897, Archibald Woods Paull, Jr. reorganized his father's company as The Wheeling Stamping Company. The company continued to prosper and by 1899 the production was up to 192,000 units. Many lanterns made by the Wheeling company have the name Paull's embossed on them.

In 1946 the Lantern Division of the Wheeling Stamping Company was purchased by the R. E. Dietz Company. The remainder of the company continued to operate in Wheeling.

The Leader has an unusually large fount. The plastic fount fill caps were unique to the Wheeling Stamping Company. See Paull's Regal in Plate 7.34 for one of Wheeling Stamping Company's hot blast lanterns.

Variations:

The Leader was available in green paint or tin finish and the globe could be red or clear.

Plate 7.33
Wheeling Leader, 1920-1946, 15.0 in.(38.1 cm) 7.75 in.(19.7 cm) $20-$30 USD

PAULL'S
LEADER
LANTERNS
& LAMP
BURNERS
SINCE 1877

Plate 7.33a

Wheeling Regal

Plate 7.34 **Description**:

The Paull's Regal is an all steel, hot blast lantern of ordinary design. The burner is a steel, No. 1, with a 0.625 inch (1.6 cm) wick. The lantern uses a No. 0 globe and it has an inside lift and a large, one piece, aluminum fount cap. The original finish is red paint.

Markings: (embossed on fount)
- REGAL -
No 0

Patent Information: (embossed on the crown)
Pat.
JULY 29th 1890
Pat.
JUNE 30th 1903

Remarks:

The Nail City Lantern Company of Wheeling, West Virginia was started in 1877 by Archibald Woods Paull, Sr. Nail City produced a variety of dead flame railroad style lanterns. By 1895, production had reached 184,500 lanterns. In 1897, Archibald Woods Paull, Jr. reorganized his father's company as The Wheeling Stamping Company. The company continued to prosper and by 1899 the production was up to 192,000 units. Lanterns made by the Wheeling company have the name Paull's embossed on them.

In 1946 the Lantern Division of the Wheeling Stamping Company was purchased by the R. E. Dietz Company. The remainder of the company continues to operate in Wheeling to this day.

The dies used to form this example must have been in use for a very long time. It makes little sense to apply patent dates that are long out of effect. By the time this lantern was built, the patents cited on the crown were completely out-of-date.

Plastic is an unusual material to find used for fount caps and it seems to be unique to the Wheeling Stamping Co. A generic Corning globe or other plain No. 0 globe would be appropriate for this lantern. The Corning Glass Company of Corning, New York, was a major producer of lantern globes. Corning invented heat resistant glass for railroad lanterns in 1909. The heat resistant glass evolved into the Corning "Pyrex" we know today. Corning globes are marked with a large "C" with a "NX" inside.

The Regal and other Wheeling lanterns (Plate 7.33) are very plain looking with no innovations to note. They are easily located and commonly collected.

Variations:

Wheeling Stamping made a number of hot and cold blast lanterns including the Paull's No. 0, which is nearly identical to the Regal but with a smaller fount.

Plate 7.34
Wheeling Regal, 1920-1946, 13.5 in.(34.3 cm)
6.75 in.(17.0 cm) $10-$20 USD

PAUL'S LANTERNS

WHEELING, W. VA

Plate 7.34a

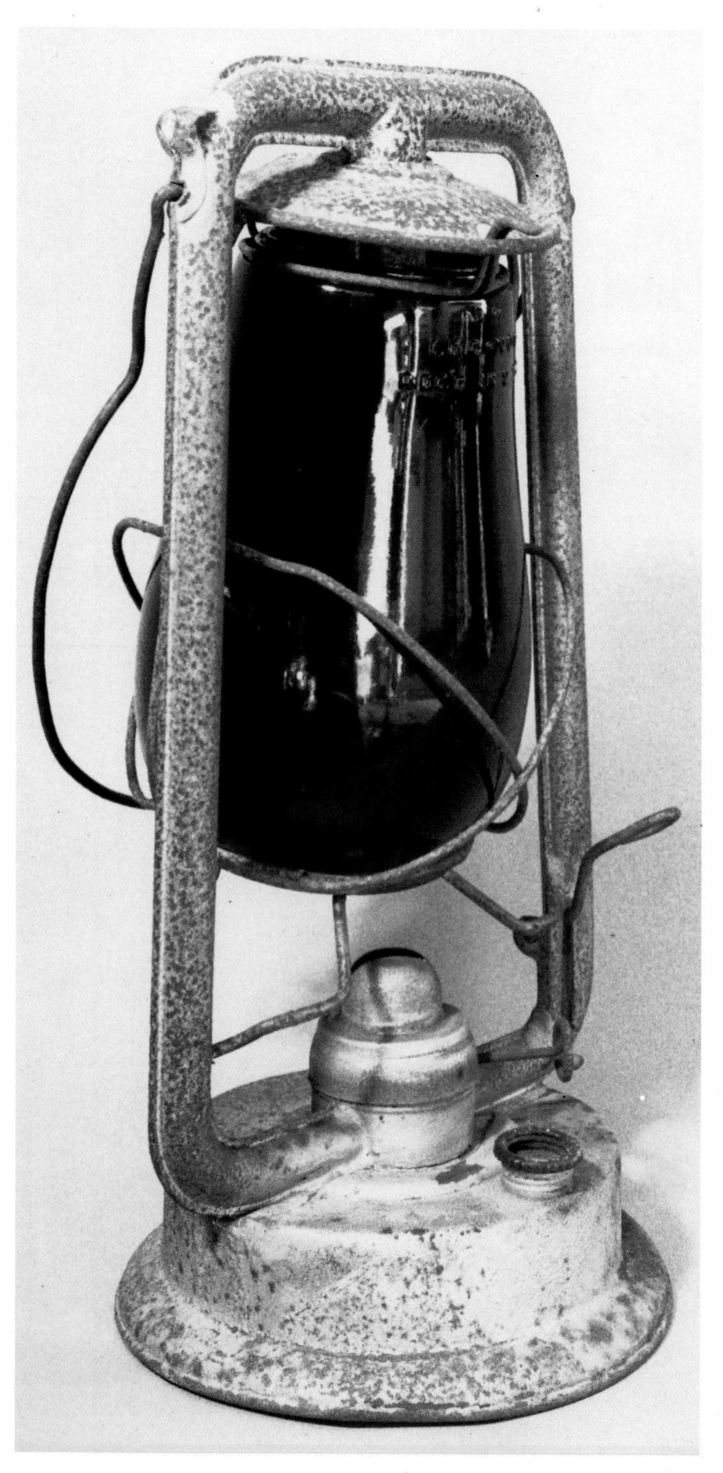

Plate 7.34b
The Regal, before restoration, is shown in its open position.

Perkins Perko

Plate 7.35 **Description**:

This Perkins Perko is a solid brass, dead flame lantern, made for use as the mast head or bow marker on a boat. This model has a clear, 180°, fresnel lens but 90°, and 360° versions were also available. The lens is 4 inches by 4 inches (10.2 cm). The lantern has a spring loaded, drop fount with an internal adjust, brass burner using a 0.625 inch (1.6 cm) wick. This specimen is polished with a clear lacquer finish and the glass is crizzled.

Markings: (on the back)

PERKO
PERKINS M. L. CO. BKLYN N. Y.
The World's Largest
Manufacturer of its Kind
(graphic of a searchlight)

(on the vortex kerosene burner)

PERKO
U.S.A.

Patent Information: (none)

Remarks:

Frederick Perkins started Perkins Marine Lamp Corporation in 1916 to produce brass and hot galvanized iron lanterns and other ship fixtures. The World Standard Deck lantern of Plate 6.4 was made by Perkins around this time. "The World's Largest Manufacturer of its Kind" is a big slogan but, judging by the number of Perkin's lanterns in the antique market, the Perkins Marine Lighting Company was indeed a big operation.

The Great Depression caused Perkins to reorganize and change the corporation name to Perkins Marine Lamp and Hardware. The company prospered through the 1930s, 40s, and 50s. The corporation moved to Miami, Florida in 1960 and the name again changed to Perko, inc.

The complete set of nautical clearance lights include this 180° light for the mast, two 90° lights for the port and starboard side, and a 360° anchor or stern light.

The Perkins company also made some railroad style lanterns with the name VACLITE, but no known tubular types.

Solid brass or heavy galvanized steel is necessary for use around boats and salt water.

Variations:

This Perko light is smallest size, suitable for boats up to 25 feet (7.6 meters) only. Class 2 boats range from 26 to 39 feet (7.9 to 11.9 meters) and class 3 boats range from 40 to 65 feet (12.2 to 19.8 meters). The sizes of boat lights began at 10 inches (26.7 cm) tall and continue in 2 inch (5.1 cm) increments, up to the English Channel size of 21 inches (53.3 cm). The larger the boat, the larger the lights required by law. All are available in brass or galvanized iron.

Plate 7.35
Perkins Perko, 1920-1930, 10.5 in.(26.7 cm) 5.25 in.(13.3 cm) $210-$280 USD

Plate 7.35a
Brass label on the back of this Perkin's lantern was first used in 1920.

Chapter 8

All Machine Construction

This chapter is devoted to modern lanterns that are designed to be assembled with the minimum amount of hand labor. This usually means the air tubes are attached to the fount with a machine crimp rather than by solder. Machine made lanterns still have many hand operations but the soldering processes have been mostly eliminated. The elimination of manual tasks speeds up production and reduced the cost to the consumer.

The railroad styles were the first to eliminate solder by electrically welding wires to each other and to the base. This process saves time because the weld is almost instantaneous. Faster construction means a reduced cost that can be passed on to the consumer. Improved value is what made the 1939 Monarch the most successful kerosene barn lantern ever produced.

Table 8.1, Lanterns Made by Machine (sorted by date)

PLATE	LANTERN MAKER/NAME	DATE	TYPE/STYLE	LANTERN MAKER	OVER-ALL HEIGHT	OVER-ALL WIDTH	GLOBE TYPE	WICK(in.)
8.1	Adlake Kero Railroad	1930-P	dead flame	Adams & Westlake	9.5 in. (24.1 cm)	8.0 in.(20.3 cm)	kero	.625 & .875
8.2	Handlan St. Louis	1935-1960s	dead flame	Handlan St. Louis	9.5 in. (24.1 cm)	6.25 in.(15.9 cm)	special	0.25
8.3	Dietz Monarch (1936)	1936-1962	hot blast	R. E. Dietz Co.	13.5 in. (34.3 cm)	6.5 in.(16.5 cm)	LOC-NOB	.625
8.4	Embury Air Pilot No. 2	1936-1952	cold blast	Embury Mfg Co.	13.25 in. (33.7 cm)	7.5 in.(19.0 cm)	short	.875
8.5	Embury Little Air Pilot	1936-1952	cold blast	Embury Mfg	13.25 in. (33.7 cm)	7.5 in.(19.0 cm)	wizard	.625
8.6	Embury No. 0 Air Pilot	1936-1952	hot blast	Embury Mfg. Co.	13.0 in. (33.0 cm)	6.5 in.(16.5 cm)	No. 0	.625
8.7	Dietz Little Wizard (1938)	1938-1962	cold blast	R. E. Dietz Co.	11.5 in. (29.2 cm)	6.5 in.(16.5 cm)	littlewizard	.625
8.8	Dietz Blizzard (1939)	1939-1962	cold blast	R. E. Dietz Co.	14.25 in. (36.2 cm)	7.5 in.(19.0 cm)	LOC-NOB	.875
8.9	Dietz Comet (No. 50)	1939-P	cold blast	R. E. Dietz Ltd.	8.5 in. (21.6 cm)	4.25 in.(10.8 cm)	comet	0.38
8.10	Dietz D-Lite (1939)	1939-1962	cold blast	R. E. Dietz Co.	13.0 in. (33.0 cm)	7.75 in.(19.7 cm)	short	.875
8.11	Dietz No. 100	1939-1962	cold blast	R. E. Dietz Co.	12.25 in. (31.0 cm)	7.5 in.(19.0 cm)	wizard	0.38
8.12	Embury/Dietz Traffic Gard	1945-1962	dead flame	Embury & Dietz	8.25 in. (21.0 cm)	7.25 in.(18.4 cm)	special	0.25
8.13	Dietz Torch	1950-1960	dead flame	R. E. Dietz Co.	8.0 in. (20.3 cm)	8.0 in.(20.3 cm)	none	0.75
8.14	Dietz Air PilotNo. 8	1953-P	cold blast	R. E. Dietz Ltd.	13.75 in. (34.9 cm)	7.75 in.(19.7 cm)	No. 0	.875
8.15	Dietz Junior No. 20	1956-P	cold blast	R. E. Dietz Ltd.	12.25 in. (31.1 cm)	5.25 in.(13.3 cm)	junior	.625
8.16	Dietz Little Wizard No. 1	1956-P	cold blast	R. E. Dietz Ltd.	12.0 in. (30.5 cm)	7.75 in.(19.7 cm)	wizard	.625
8.17	Dietz Blizzard No. 80	1956-P	cold blast	R. E. Dietz Ltd.	15.0 in. (38.1 cm)	7.75 in.(19.7 cm)	No. 0	.875
8.18	Dietz D-Lite No. 90	1956-P	cold blast	R. E. Dietz Ltd.	13.5 in. (34.3 cm)	7.75 in.(19.7 cm)	Short	.875
8.19	Dietz Monarch No. 10	1956-P	hot blast	R. E. Dietz Ltd.	13.75 in. (34.9 cm)	5.25 in.(13.3 cm)	No. 0	.625
8.20	Dietz The Original	1976-P	cold blast	R. E. Dietz Ltd.	10.25 in. (26.0 cm)	4.5 in.(11.4 cm)	special	0.50

-P (to the present) indicates the lantern is still available

Adlake Kero

Plate 8.1 **Description**:

The Kero is a dead flame, single guard, wire bottom, railroad style lantern with a steel pot and burner. The 300 burner uses a 0.625 inch (1.6 cm) wick and the 400 uses a 0.875 inch (2.2 cm) wick. The wick adjust is a bent wire type. The balance of the lantern is crimped, riveted, and electrically welded steel. The pot and burner drop in from the top. The Kero lantern uses a kero globe available in clear, red, yellow, green, or blue. The Kero has a bail lock and the finish is tin over steel.

Markings: (embossed on crown)
ADLAKE PC KERO
(embossed on globe)
ADLAKE KERO
(embossed on pot)
No. 400 USE LONG TIME BURNING OIL ONLY
(stamped on burner)
ADLAKE 400
(embossed on bottom)
ADLAKE KERO
USA CANADA PATENTED

Patent Information: (none)

Many Kero lanterns produced from 1930 to 1965 are marked with a production date on the bottom.

Plate 8.1
Adlake Kero Railroad, 1930, 9.5 in.(24.1 cm) 8.0 in.(20.3 cm) $50-$200

Remarks:

The Adlake Kero may be the most common railroad style lantern in North America. There are more than 100 documented user markings including some non-railroad companies like Lionel toy trains.

The Kero's design uses a minimum number of parts and it has no solder so the cost was lower than the competition.

This Kero lantern is marked with the Penn Central "snakes in love" logo. A good source of detailed railroad lantern information can be found in The Illustrated Encyclopedia of Railroad Lighting by Richard C. Barrett.

Variations:

The details of the Adlake Kero changed over the years and lots of options are available. Besides globe colors there are blinders to control the direction of the light, a heavy base for stability, iron mounting hardware, and bails made of wire, metal, wood, or fiber.

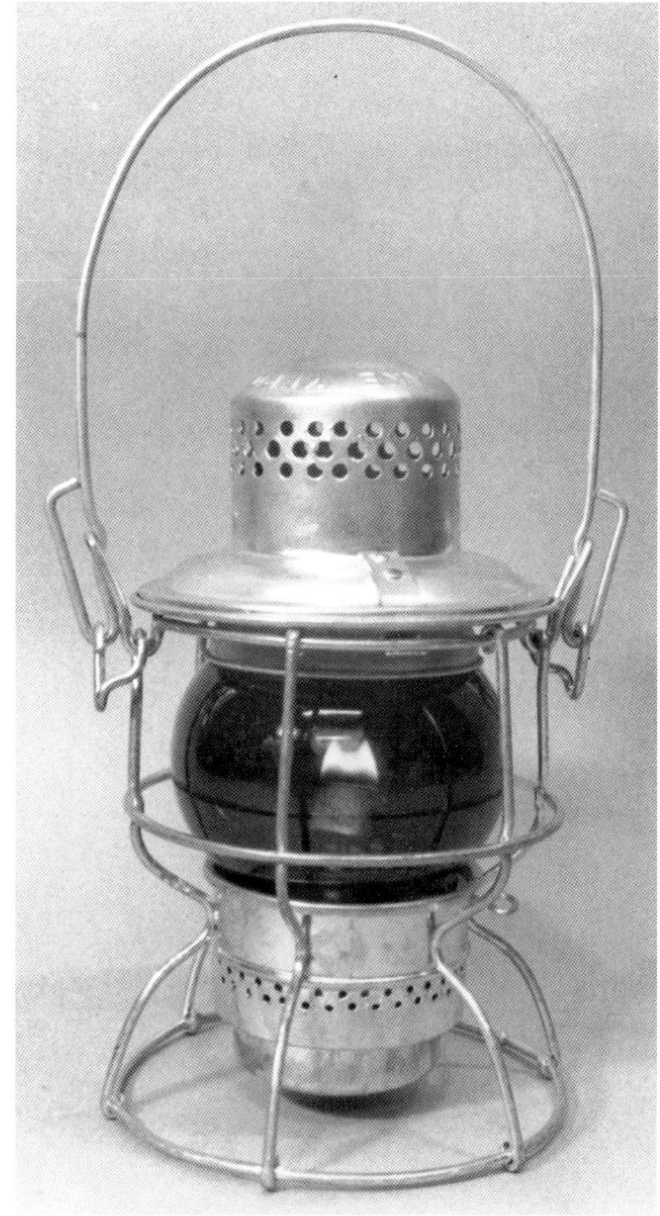

Plate 8.1a
Adams and Westlake is still in business and still makes products for the railroad industry. http://www.adlake.com. From the collection of Scott E. Schifer.

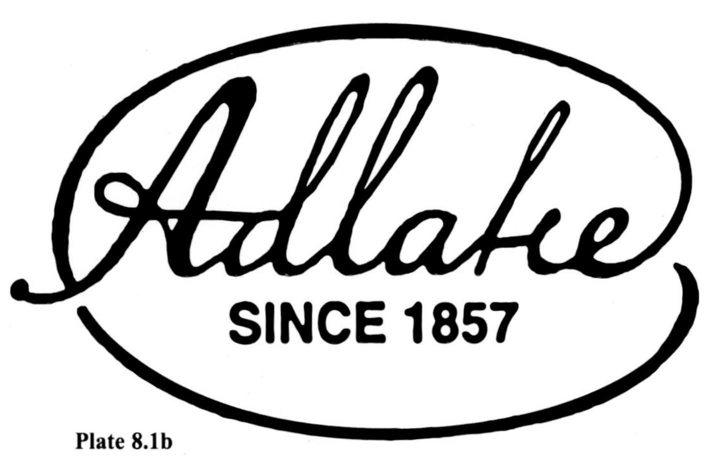

Plate 8.1b

Handlan St. Louis

Plate 8.2 **Description**:

This contractor's, dead flame lantern has a steel, No. 101 burner with a 0.25 inch (0.64 cm) round, felt wick. The balance of the lantern is electrically welded, soldered and crimped steel. The fount is not removable and serves as the base of the lantern. The fill cap is a large, one piece steel stamping. This lantern uses a Handlan No. 364 globe and has a flip-up crown. Original finish is tin plate over the steel.

Plate 8.2
Handlan St. Louis, 1935-1960s, 9.5 in.(24.1 cm) 6.25 in.(15.9 cm) $20-$30

Markings: (on crown)
HANDLAN
ST. LOUIS U.S.A.
(on globe)
HANDLAN
ST. LOUIS
INC.
(on fount or crown)
PROPERTY OF GAS CO.
Patent Information: (on burner plate)
PATENT NO.
1959128
(corresponds to around 1935)
Remarks:

This lantern's unfortunate similarity to a railroad style lantern has made it one of the most misidentified lanterns of all time. Usually found with the utility's name embossed on the crown or fount, the Handlan is tagged and priced as a "railroad" lantern rather than a contractor's lantern.

This Handlan became the last descendant of the great Handlan-Buck line of lanterns.

In 1856, Myron. M. Buck opened a factory in St. Louis, to manufacture railroad car fittings. An employee, Alexander Handlan, bought out Mr. Buck in 1901 and eventually changed the company name to Handlan-Buck. The company was a very successful producer of signal lanterns and even supplemented their catalogue with some Dietz tubular lanterns.

Alexander. H. Handlan, Jr. took over the company in 1921 and the company continued to sell lanterns and parts into the 1970s.

This lantern is sometimes marked with railroad company names. Compare to the earlier Handlan brakemen's lantern of Plate 7.9.

The unique No. 364 globe is 2.5 inches (6.4 cm) diameter at the top, 3.25 inches (8.3 cm) at the bottom, and 4.5 inches (11.4 cm) tall. It is not interchangeable with any other standard globe.

The Handlan catalogue lists the following wick sizes:

#00 for P&A Nutmeg — *3/8 in. flat*
#0 or E — *1/2 in. flat*
#1 or A — *0.625 in. flat*
#2 or B — *1 in. flat*
#3 or D — *1-1/2 in. flat*
longtime gray (felt) — *1/4 in. round*
TORCH WICKS
1/2 in. x 12 in.
3/4 in. x 18 in.

Variations:

The No. 364 globe could be ordered in white (clear), green, yellow, blue, and red. The No. 101 slip in burner could be ordered for the #00, #0, #1, #2, or longtime gray wick.

Sometime after the war, the crown was lowered by 0.5 inch (1.3 cm) and the fount was crimped as shown on the right of Plate 8.2. The newer version usually has a ruby fresnel globe rather than the old railroad style. The fresnel globe has the same dimensions as the No. 364.

Terne plate was used in place of tin during World War Two because tin was rationed. Handlan St. Louis

Plate 8.2a
The later model Handlan, with a crimp fount, was popular with utilities.

Dietz Monarch (1936)

Plate 8.3 **Description**:

This all steel, streamline, hot blast lantern has a large, two piece fount cap embossed with the Dietz logo. It has a LOC-NOB globe, inside lift, all steel, No. 1, domed burner, with a 0.625 inch (1.6 cm) wick and bent wire wick adjust. The air tubes are uniquely rounded at the top and stamp embossed for strength. Machine crimp construction is used throughout with just a touch of solder at the tube joint near the top.

Markings: (embossed on crown)
DIETZ MONARCH NEW YORK, N.Y. U.S.A.
(embossed on globe)
DIETZ (logo)
FITZALL
N. Y. U. S. A.
H18
LOC-NOB
REG'D U.S. PAT. OFF.
(embossed on fount)
DIETZ N.Y. U.S.A. MONARCH

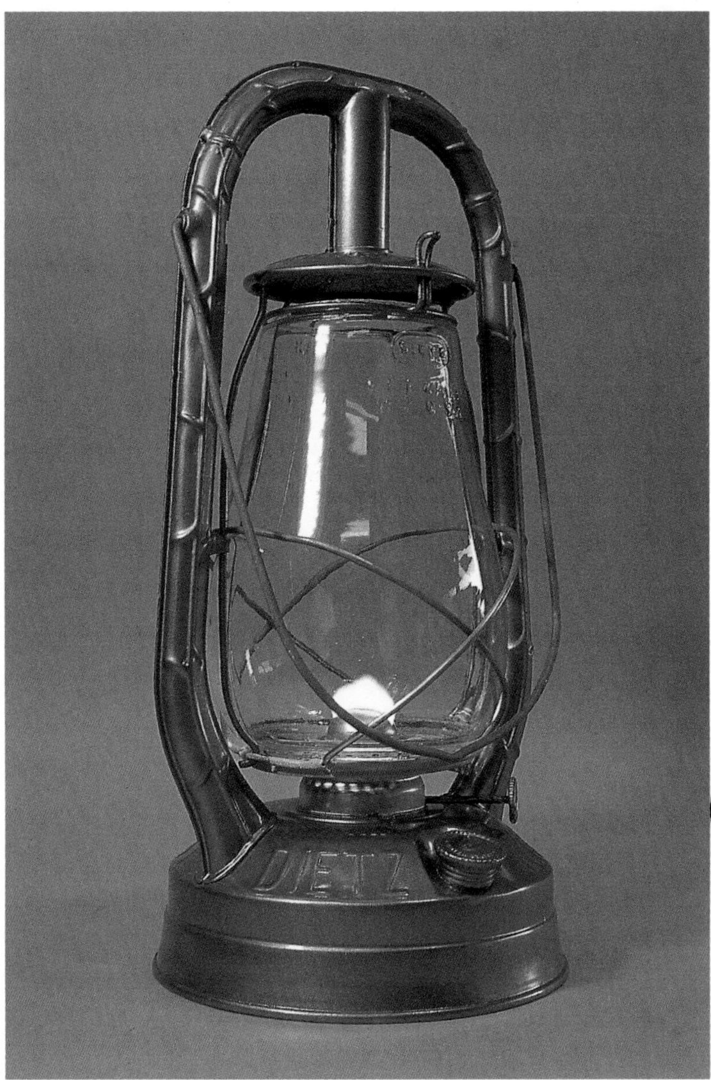

Plate 8.3
Dietz Monarch (1936), 1936-1962, 13.5 in.(34.3 cm) 6.5 in.(16.5 cm) $10-$20

This lantern is commonly seen and easy to identify by its unusual, rounded side tubes. Stamped entirely by machine it was very inexpensive which made the Monarch, with a ruby globe, very popular with municipal street works and the movie industry.

The correct globe is the FITZALL or LOC-NOB No. 0 globes that fit snugly (locks) into the globe guard wires.

The correct finish for this lantern is a light metallic blue.

Variations:

This lantern is probably the most successful hot blast lantern ever produced. The Monarch was produced by special order in large numbers. Many of these special order Monarchs are embossed on the crown or fount with the owners name; DWP, Street Dept. Etc. The standard color was metallic blue but a Monarch could be special ordered in another color. The globe could be ordered in clear, or ruby.

Compare this lantern to the 1905 Monarch of Plate 6.18 and the 1913 version in Plate 7.19.

Plate 8.3a

Patent Information: (on side tube)

PAT. APD. FOR
S - 10 - 49

Remarks:

Designed in 1936, the Dietz Monarch was the first lantern to have the Art Deco restyling. It is noteworthy that the Deco style of the 1930s spilled over into the styling of lowly kerosene lantern. Note the tubes no longer just carry hot air to the burner but arch over the crown, flow around the globe, and angle into the domed fount. The ridges on the tubes are arched and angled like the decoration on the Empire State Building. The fount is flared and stepped like base of a marble statue. The contemporary styling sets the Dietz streamline lanterns apart from their klunky, chunky competitors.

Plate 8.3b
The Monarch family photo: 1900, left, 1913, and 1936, right.

Embury No. 2 Air Pilot

Plate 8.4 **Description**:

The Embury No. 2 Air Pilot is an elegant, all steel, cold blast kerosene lantern. It has an off center, two piece fount cap, and a No. 2, steel, 0.875 inch (2.2 cm), rising cone burner. The No. 2 Air Pilot has an inside lift and the body is painted Embury metallic green. A short globe is used on the No. 2.

Markings: (embossed on Embury globe)

No. 30
-E-
USA

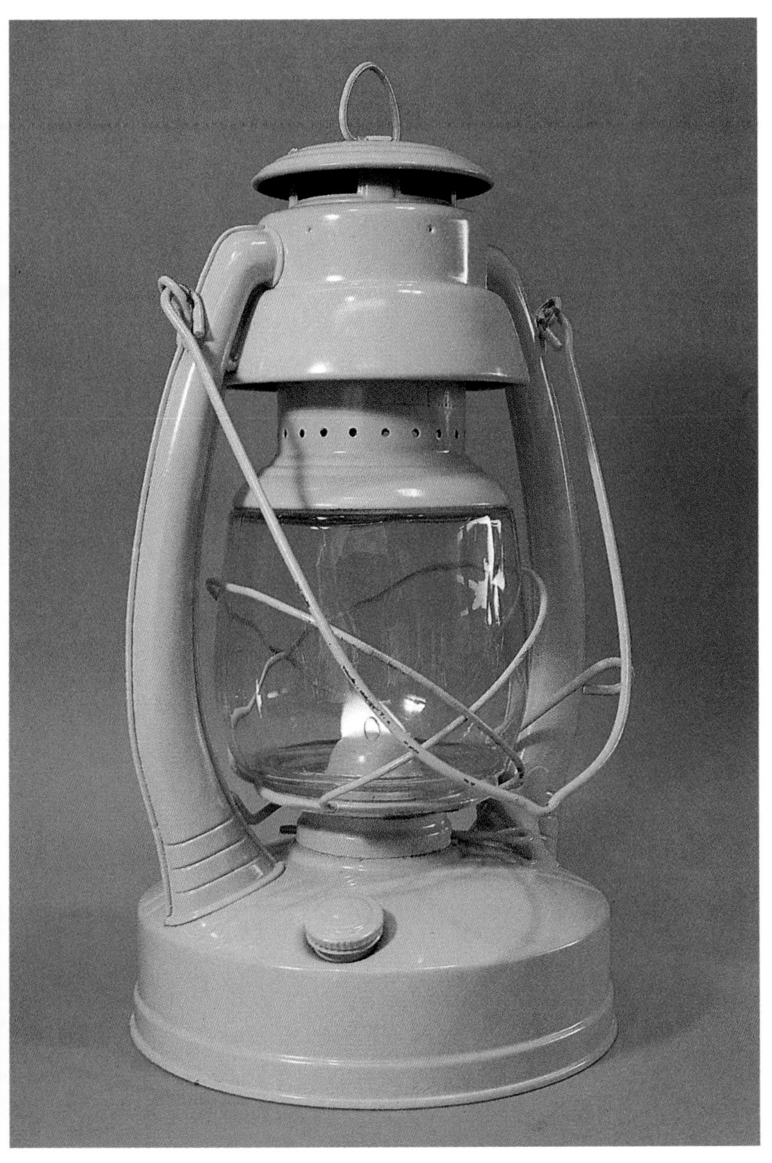

Plate 8.4
Embury Air Pilot No. 2, 1936-1952, 13.25 in.(33.7 cm) 7.5 in.(19.0 cm) $20-$30

The Standard Embury marking are:

EMBURY MFG. CO.
WARSAW, NEW YORK, U.S.A.
NO. 2 AIR PILOT

(this example: on fount - see remarks)

ELGIN

Patent Information: (on lift support)

PAT'S
1885517 (1932)
2054878 (1936)
D-111578

Remarks:

Embury was quick to follow the Dietz restyling with an even more radical update of their lantern line. With its Art Deco lines, the No. 2 Air Pilot is one of the most stylish lanterns ever produced. The flair of the bell and fount flow with the curve of the smooth air tubes. The accent lines embossed on the fount and tubes are not overpowering, but break the monotony. This lantern's bold statement makes it one of the all time classics.

Embury embossed Air Pilots with the names of different jobbers, distributors, utilities and municipalities. This example is marked simply ELGIN. Elgin could be the name of the city that special ordered this lantern; Elgin, Illinois.

All No. 2 Air Pilots have a star embossed on the bottom.

Variations:

Because of its modern design the No. 2 Air Pilot was very popular with distributors. It must be the lantern with the most identities. Everyone wanted their name on it. The Air Pilot has been found marked:

SHAPLEIGH HARDWARE CO. ST. LOUIS, U.S.A. NORLEIGH DIAMOND
WARD'S BETTER LANTERN
VAN CAMP No. 180 AIR PILOT
HERCULES
GAMBLE'S ARTISAN
BELKNAP HWD. & MFG. CO. BLUE GRASS MADE IN U.S.A.
COAST-TO-COAST STORES MASTERCRAFT NO. 2

Plate 8.4a

Embury Little Air Pilot

Plate 8.5 **Description**:

The Embury Little Air Pilot is a nicely style, all steel, cold blast lantern using a little wizard size globe. It has a steel, No. 1, rising cone burner, 0.625 inch (1.6 cm) wick, Embury loop wire wick adjust, curved air tubes, off-center fill with a two piece, steel cap, and an inside lift. Construction is all machine crimp. Finish is Embury green metallic paint. This contractor's lantern has a ruby globe but clear globes were also available. All were made in Warsaw, New York.

Markings: (embossed on fount)

EMBURY No 350 LITTLE AIR PILOT
EMBURY MFG CO.
WARSAW - NY - USA

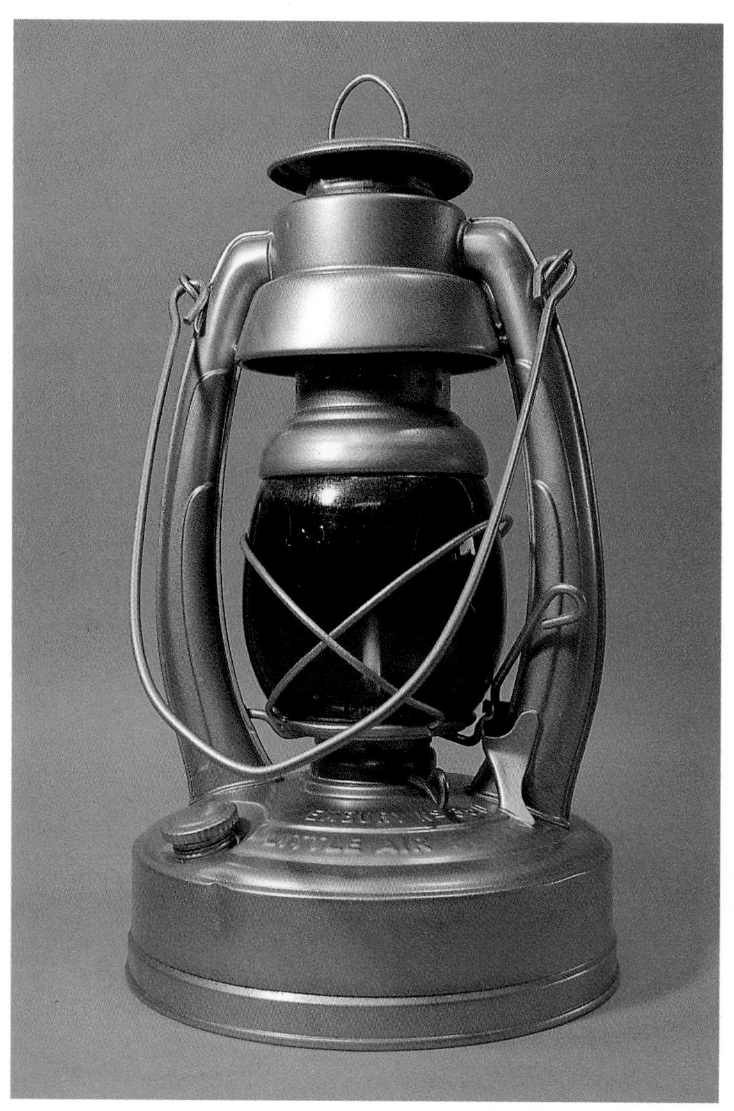

Plate 8.5
Embury Little Air Pilot, 1936-1952, 13.25 in.(33.7 cm) 7.5 in.(19.0 cm) $20-$30

(embossed on globe)

10

Patent Information: (stamped on lift support)

PATS
1985517 (1935)
2054878 (1936)
D-111578

Remarks:

The elegantly styled Little Air Pilot was manufactured from 1945 to 1962. The Embury Manufacturing Company was incorporated in 1908 by William C. Embury and by 1923 it had 175 employees and sales of 900,000 units. In December 1930, Embury purchased the Defiance Lantern and Stamping Company. Bill Embury Sr. passed away on June 11, 1943 and the Embury company was dissolved by Phil, Fred, and Bill Jr. on the last day of 1952. The R. E. Dietz Company purchased the tools and equipment in 1953 and continued to make some Embury designs until 1962.

Besides its artistic styling the Little Air Pilot's fount has an unusual construction feature. The fount is made of three pieces of metal. The bottom half telescopes into the top half, making the sides of the fount double wall. Was this added strength a selling feature?

The Little Air Pilot is a restyled Little Supreme (Plate 7.31) used by contractors and street departments as safety lights around construction projects. They are not known to be purchased or used by railroads. Because of the red globe, unknowledgeable and unscrupulous persons have misrepresented contractor's lanterns as more popular railroad lanterns.

Many of these lanterns were purchased in large numbers by municipalities. In quantity, lanterns could be special ordered with the city or utility name embossed in metal. It is not unusual to see Embury lanterns factory embossed with STREET DEPT., D W & P, or PROPERTY OF (your city). In fact it is unusual to find an Embury lantern marked for Embury Manufacturing Company.

Variations:

Besides the embossed customer's name this lantern could be special ordered in a variety of paint colors. A clear globe was also available.

Plate 8.5a
Note the Art Deco text on the Little Air Pilot's fount.

Embury No. 0 Air Pilot

Plate 8.6 **Description**:

The Embury No. 0 Air Pilot is an all steel, hot blast, kerosene lantern. It has an off center, two piece, large, steel fount cap, and a No. 1, steel, 0.625 inch (1.6 cm), domed burner. This Air Pilot has an unusual rear lift and all were painted with Embury Metallic green paint. The globe is a No. 0 size, ruby, with an E for Embury embossed on it. The bottom is embossed with a star just like the No. 2 Air Pilot. From the collection of Scott E. Schifer.

Markings: (embossed on Embury globe)

No. 20
-E-

(on fount)

EMBURY NO. 0 AIR PILOT
D. W. & P.

Patent Information: (none)

Remarks:

The Embury Manufacturing Company of Rochester, New York was incorporated November 27, 1908 by William C. Embury. The company designed and produced a variety of tubular and railroad style lanterns. The factory moved to Warsaw, New York in 1911, where it remained until the doors were closed on December 31, 1952.

There are several lanterns in the Embury line named Air Pilot, which must have created confusion. There were so many cities, counties, utilities, and contractors purchasing Air Pilots of one type or another from Embury, that Dietz continued the name after they purchased the equipment in 1953.

It was very common for the Embury company to build a special run for jobbers, distributors, utilities and municipalities. These lanterns can be found with a wide variety of markings. This example is marked for a generic Department of Water and Power (D.W.& P.).

Compare this hot blast lantern with the cold blast No. 2 Air Pilot of Plate 8.4.

Variations:

This lantern, and all the other Embury lanterns, were available with ruby or clear globes.

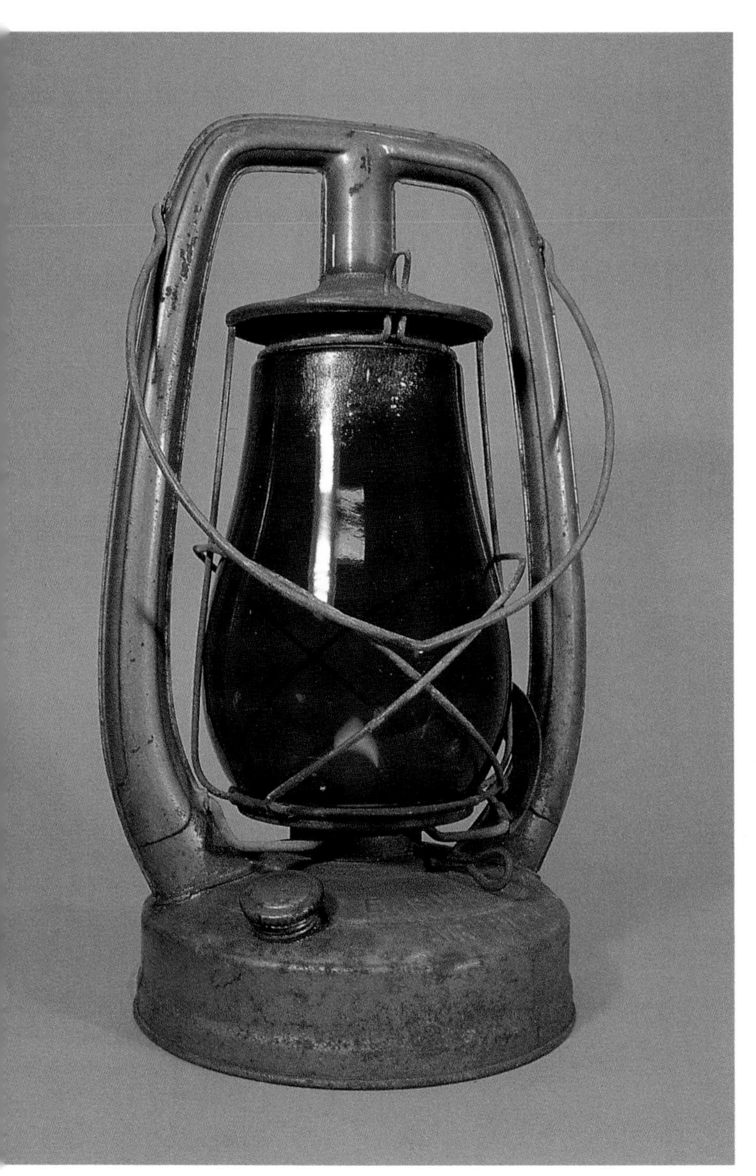

Plate 8.6
Embury No. 0 Air Pilot, 1936-1952, 13.0 in.(33.0 cm) 6.5 in.(16.5 cm) $20-$30

Plate 8.6a
A perfect Embury globe is unusual to find perhaps because many suffered from diseased glass as this one does.

Plate 8.6b
The Department of Water and Power purchased this lantern for street safety.

Variations:

This lantern, with its variations, is the most successful contractor's lantern ever produced. The Little Wizard was made in large numbers, with the name of the owner embossed on the chimney flare or on the fount. Wizard lanterns can be found with name of a city street department, Gas Company, phone company, DWP, etc., stamped in the metal by the factory in Syracuse, New York.

The standard color was metallic blue but a Little Wizard could be special ordered in gray, red, or galvanized. The globe could be ordered in clear, although red is more commonly found.

Compare this lantern to the other lanterns that use the Little Wizard globe; No. 100 of Plate 8.11 and the Little Giant in Plate 7.32.

Also note that the 1920 Little Giant dies were used to produce the imported Little Wizard No. 1 in the 70s, 80s, and 90s because the 1938 Wizard dies simply wore out.

Dietz Little Wizard

Plate 8.7 **Description**:

This all steel cold blast lantern will burn for two days due to its very large fount. It has a clear, little wizard size globe, inside lift, all steel, No. 1, rising cone burner, with a 0.625 inch (1.6 cm) wick and bent wire wick adjust. The large, two piece fount cap is embossed with the Dietz logo. The air tubes flare out from the dome fount, are rounded at the top and stamp embossed for strength. Machine crimp construction is used throughout.

Markings: (on crown)

DIETZ LITTLE WIZARD N. Y. U.S.A.

(embossed on globe)

DIETZ (logo)
LITTLE WIZARD
N. Y. U.S.A.

(embossed on fount)

LITTLE WIZARD
DIETZ N. Y. U.S.A.
PATENT NUMBERS:

(on side tube)

1338911
1795542
1892292
2062051
2079125
S - ? - 52

Remarks:

First issued in 1938, the Dietz Little Wizard was the first cold blast lantern to get the Art Deco treatment. It may also be the first non-railroad style lantern to eliminate solder.

This lantern is commonly seen, and easy to identify, by its short stature and broad fount. The large fount allowed extended operation without maintenance so the Little Wizard, with a ruby globe, was very popular with municipalities and utilities.

The correct finish for this lantern is Dietz metallic blue with a clear or ruby, Little Wizard, LOC-NOB globe. Note the patent numbers and date stamped into the side tube. The numbers indicate 1920, 1930, 1932, 1936, 1937, and 1952.

The Little Wizard sold new for $8.50 in 1975 and $19.04 in 1996.

Plate 8.7
Dietz Little Wizard (1938), 1938-1962, 11.5 in.(29.2 cm) 6.5 in.(16.5 cm) $10-$20

Plate 8.7a
The original 1920 Little Wizard is designed for contractor's use.

Dietz Blizzard No. 2

Plate 8.8 **Description**:

The Dietz Blizzard is a large, all steel, cold blast lantern that has a two piece fount cap embossed with the Dietz logo. It has a LOC-NOB globe, inside lift, an all steel No. 2, domed burner, with a 0.875 inch (2.2 cm) wick and bent wire wick adjust. The air tubes are machine embossed for strength. Machine crimp construction is used throughout. All Dietz lanterns made after the close of the New York city factory in 1931, were made in Syracuse, New York.

Markings: (embossed on crown)
DIETZ (logo)
(embossed on globe)
DIETZ (logo)

Plate 8.7b
The 1938 Little Wizard.

Plate 8.8
Dietz Blizzard (1939), 1939-1962, 14.25 in.(36.2 cm) 7.5 in.(19.0 cm) $10-$20

BLIZZARD
N. Y. U. S. A.
H18
LOC-NOB
REG'D U.S. PAT. OFF.
(embossed on fount)
DIETZ N. Y. U. S. A.
NO. 2 BLIZZARD

Patent Information: (embossed on side tube)
PATENTS
1338911
1795542
1892292
D101113
2062051
2079125
???-??-4?

Remarks:

When it was the Blizzard's turn to be restyled, Dietz, gave it the Art Deco lines of the 1936 Monarch (Plate 8.3). The curve of the tubes is not as noticeable in a cold blast lantern but the shape of the fount is still there. The Blizzard's air tubes are now decorated with the structural ridges. The globe is embossed with the LOC-NOB data plus the name BLIZZARD. The correct finish is tin or Dietz metallic blue paint.

The Dietz Blizzard produces one of the brightest lights by using the largest globe (No. 0) and wick (No. 2) available in a hand lantern. The large wick burned fuel too fast for contractor use so the Blizzard was marketed to home and industrial users.

The name Blizzard is taken from a successful line of lanterns first built by Dietz in 1898 (see Plate 6.15).

The patent numbers on the tube indicate 1920, 1930, 1932, 1936, 1939, and 1954.

Variations:

This example is unpainted tin plate but the classic Dietz metallic blue paint and special order colors were also available.

Plate 8.8a
Late Dietz slogan.

Dietz Comet

Plate 8.9 **Description**:

The Dietz Comet is a tiny, all steel, cold blast lantern with a large, two piece, steel fount cap and a No. 0 burner (domed until 1956, rising cone after) with a 0.38 inch (1 cm) wick. The Comet has an inside lift, high dome fount, and all machine crimp construction. The air tubes are machine embossed for strength. The U.S. and Hong Kong versions are painted red.

Plate 8.9
Dietz Comet, (No. 50), 1939, 8.5 in.(21.6 cm) 4.25 in.(10.8 cm) $20-$30

Markings: (embossed on crown, all models)
DIETZ (in oval)
(on globe, 1939-1956)
H-15
DIETZ (logo)
SYRACUSE, N.Y. U.S.A.
COMET
REG'D U.S. PAT. OFF.

(on fount, 1939-1956)
DIETZ COMET
MADE IN UNITED STATES OF AMERICA
(with full text fill cap)
(on fount, after 1956)
DIETZ No. 50
(on burner, 1939-1956)
DIETZ MADE IN
UNITED STATES OF AMERICA
(on burner dome, 1939-1956)
DIETZ COMET BURNER
MADE IN U.S. OF AMERICA
(embossed on globe and fill cap, 1956-1989)
DIETZ (in oval)
(painted on globe, after 1989)
DIETZ (in oval)
(embossed on bottom, 1956-1989)
MADE IN HONG KONG
(gold label on bottom, after 1989)
Made in the People's Republic of China

Patent Information: (embossed on side tube)
1892292
2062051
S-1-51

Remarks:

Designed in 1939, the Dietz Comet was sold overseas until 1945. The Comet tooling was updated in 1956 when production started in Hong Kong. The tooling had to be changed to remove the traditional markings, N.Y. USA, from the globe and fount. The Comet name was changed to No. 50 to be consistent with the numbers added to all the other lanterns and the bottom was marked with MADE IN HONG KONG.

The patent numbers on this example are for 1933, 1936, and September 1, 1951.

The globe used in the Comet is 2.5 inch diameter top and bottom and 2.75 inches tall. Despite the scaled down dimensions, the Comet has a good flame once it gets warmed up. Years ago, these small lanterns were approved for use by the Boy Scouts of America.

A new No. 50 Comet cost $10.51 in 1996.

Variations:

Some Comet lanterns have RANCH CRAFT on the fount and RC on the cap. After 1986 the No. 50 was available in tin, solid brass, antique bronze, red, or blue.

Plate 8.9b
The No. 50 came with several finishes including bronze.

Plate 8.9a
From left to right, 1939 New York, 1956 Hong Kong, 1986 China.

Dietz D-Lite (1939)

Plate 8.10 **Description**:

The all steel Dietz D-Lite is a cold blast lantern with the large, two piece fount cap embossed with the full Dietz text. The D-Lite has an inside lift and a high dome fount. The air tubes are machine embossed for strength. The correct globe is a "short" D-LITE LOC-NOB globe that locks into the globe guard wires. The 1939 D-Lite has a steel, domed, burner with a 0.875 inch (2.2 cm) wick and the wick adjust knob is a wire bent into a circle.

Markings: (embossed on fount and crown)
No. 2 D-LITE N.Y. U.S.A. DIETZ
(full text cap reads)
DIETZ (logo) · MADE IN ·
UNITED STATES OF AMERICA

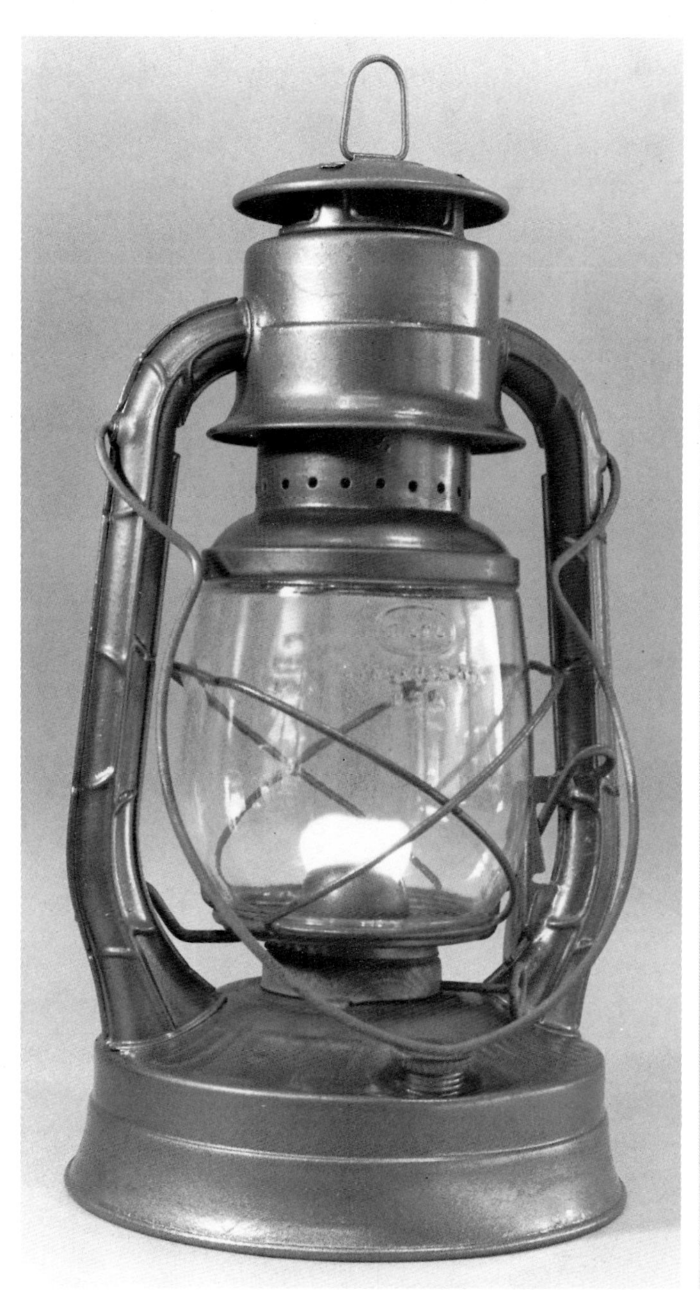

Plate 8.10
Dietz D-Lite (1939), 1939-1962, 13.0 in.(33.0 cm) 7.75 in.(19.7 cm) $20-$30

Patent Information: (on side tube)

1338911
1795542
1892292
D 19113
2062051
2071125
S -??- 49

Remarks:

The D-Lite was restyled by Dietz in 1939. Compare this lantern to the 1913 D-Lite version of Plate 7.19. The styling of this cold blast lantern was the result of the success of the first Art Deco lantern, D-Lite's older brother, the Monarch of Plate 8.3. Dietz continued to sell the older style D-Lite along side the Deco version until 1944.

This lantern is easy to identify by its short globe and rounded side tubes. Built entirely by machine, it was very popular. Note the patent information is stamped into the side tube. The numbers indicate 1920, 1930, 1932, 1936, 1937, and 1949.

When Dietz purchased the equipment and tool of the defunct Embury Manufacturing Company in 1953, Dietz needed a lantern to sell to Embury's customers who preferred Air Pilots. There must have been some economic reason for not building more Embury Air Pilots (like non-interchangeable parts) so Dietz quickly reworked the dies for the 1939 D-Lite to make the Dietz Air Pilot.

Variations:

The D-Lite was produced in large numbers. The standard color was metallic blue or tin plated. It can have a clear or ruby LOC-NOB globe.

Plate 8.10a
The rising cone burner returned when the D-Lite tooling was used to make this Dietz Air Pilot.

Dietz No. 100

Plate 8.11 **Description**:

The No. 100 is a longer burning version of the Little Wizard. It is an all steel, cold blast lantern that will burn for three days due to its very small wick and large fount. It has a red painted, Little Wizard size globe, and an inside lift. The Dietz No. 100 uses a Comet, all steel, domed, No. 0 burner, with a 0.38 inch (1.0 cm) wick and bent wire wick adjust. The large, two piece fount cap is embossed with the full Dietz text. Machine crimp construction is used throughout. The finish is a Dietz metallic blue paint or special order color.

Markings: (on crown) (nothing)
(on chimney flare)
CITY OF L. A.
(embossed on fount)
DIETZ N. Y. U.S.A.
NO. 100

Plate 8.11
Dietz No. 100, 1939-1962, 12.25 in.(31.0 cm) 7.5 in.(19.0 cm) $10-$20

(embossed on burner)
DIETZ COMET BURNER
MADE IN U.S. OF AMERICA

Patent Information: (none)

Remarks:

This lantern was developed as a response to the contractor's need for a longer burning street lantern. The Dietz No. 100 is a modified version of the Little Wizard (Plate 8.7), first issued in 1938. The No. 100 has a smaller wick for even longer burning.

This No. 100 is commonly found and easy to identify by its short stature and broad fount. The large fount allowed it to burn for up to three days without maintenance so it was very popular with municipalities and utilities.

The unique, smooth, clear, Little Wizard size, globe is painted red. There is no marking or knobs and the red paint will fade to pink. There is always a patch left clear for viewing the flame which is sometime a band as shown or a round spot. This specimen is rusty and dented.

This lantern is factory made for the City of Los Angeles. It would be nearly impossible for a collector to find one of these in Florida. The value of a Chicago lantern in the Chicago market is stable but the value of the same lantern in another part of the country is volatile. Collectors will pay extra for a lantern from a small town (small quantity) or a distant city (small local supply). Railroad lantern collectors have had to deal with the problem of lanterns marked for local railroads. Lanterns marked for out-of-state railroads can bring a higher price only when that special buyer appears.

Compare this lantern to the other lanterns that use the Little Wizard globe: the Little Wizard of Plate 8.7 and the Little Giant in Plate 7.32.

Variations:

This lantern, with its variations (Little Giant, Little Wizard), were very successful as contractor's lanterns. The No. 100 was produced in large numbers, often with the name of a city street department, Gas Company, phone company, or DWP, stamped in the crown, fount or chimney flare.

The standard color was metallic blue but customers would special order a No. 100 in yellow, gray, red, or just about any color so they could distinguish their lanterns from one another.

Embury (Dietz) Traffic Gard

Plate 8.12 **Description**:

This little dead flame, contractor's lantern has a steel burner with a 0.25 inch (0.64 cm) round, felt wick. The lantern is made of crimped and riveted steel. The Traffic Guard has a loop wire handle and a single wire guard. The crown is hinged and the wick adjust is a Dietz style, wire loop. This lantern is unusual in that it has a cylindrical, ruby, fresnel globe. The done fount is low and wide for stability. The original finish is red paint.

Markings: (embossed on crown)
EMBURY MANUFACTURING CO.
TRAFFIC GARD (sic)
No. 40 WARSAW NEW YORK
U.S.A.
(embossed on globe)
EMBURY No. 40
(or embossed on crown)
DIETZ TRAFFIC GARD (sic)
No. 40
SYRACUSE N.Y. - U.S.A.

Plate 8.12
Embury/Dietz Traffic Gard, 1945-1962, 8.25 in.(21.0 cm) 7.25 in.(18.4 cm) $10-$20

(embossed on globe)
DIETZ No. 40
(full text Dietz fill cap)
Patent Information: (none)
Remarks:

The Traffic Gard (sic) was originally designed and built by the Embury Manufacturing Company from 1945 to the end of 1952. When Dietz purchased the tooling from Embury in 1953, they dropped the competing Dietz Night Watch and built the more popular Embury design.

This contractor's lantern sold well because it cost much less than tubular lanterns. Dietz continued to make the Traffic Guard in Syracuse until 1962.

The Traffic Guard was advertised as burning three days on one pint of kerosene.

Fitted with a ruby globe this lantern would be placed around a work area to protect the men and equipment. The globe is 3.0 inches (7.6 cm) at the top and bottom and 3.0 inches tall.
Variations:

Traffic Guard lanterns were painted red, yellow or special order colors. Globes colors include ruby, amber, green, blue and clear. The Traffic Guard can also be found with a more conventional bail.

Plate 8.12a
The dietz Night Watch was built from 1945 to 1953 and could burn 100 hours on its 0.38 inch wick. It could take a wizard or fresnel globe and this one is painted safety orange.

Dietz Torch

Plate 8.13 **Description:**

This all steel torch is the ultimate in dead flame lanterns. It has no globe or burner but a non-adjustable, 0.75 inch (1.9 cm), round, cotton wick. A steel, twist-off wick cover is retained by a chain and another chain has a ring for hanging and used as a handle. Machine crimp construction is used. The fount has a heavy weight to keep the torch upright.

Markings: (embossed on fount)

DIETZ TORCH - MADE IN U.S.A.
USE KEROSENE ONLY

(embossed on wick cover)

DIETZ No. 750 - 751 - MADE IN U.S.A.

Patent Information: (none)

Remarks:

The design of the Dietz Torch is similar to a smudge pot. The torch is the least expensive solution for protecting a construction project at night. However, using an open flame would seem to create another hazard that would not be allowed today. When knocked over, a tubular kerosene lantern will loose its draft and smother within the burner dome. There is no such safety on these torches. The torch is shaped like a ball and many a kid must have stubbed his toe, kicking these lamps down the street.

Plate 8.13
Dietz Torch, 1950-1960, 8.0 in.(20.3 cm) 8.0 in.(20.3 cm) $10-$20

The torch is advertised to burn 50 hours on three quarts (2.84 liter) of kerosene. The large wick and lack of draft produces a smoky flame. The 0.75 inch wick is available today for patio Tiki torches.

Variations:

Available in red, black and special order colors. The No. 87 torch lacks the chain and is painted with blue enamel.

Plate 8.13a
Late Dietz logo.

Dietz Air Pilot No. 8

Plate 8.14 **Description**:

This steel with brass trim, streamline, cold blast lantern has a large, two piece fount cap embossed with the Dietz logo. It has a metric globe, inside lift, all steel No. 1 rising cone burner, with a 0.875 inch (2.2 cm) wick and bent wire wick adjust. The air tubes are machine embossed for strength. Machine crimp construction is used throughout. First made in Syracuse and later imported from Hong Kong and China.

Markings: (embossed on fount)

DIETZ NO. 8 AIR PILOT

(embossed on bell)

DIETZ

(painted on globe)

DIETZ

(gold label on bottom)

Made in the People's Republic of China

Patent Information: (none)

Remarks:

The name Air Pilot is taken from a successful line of lanterns built by Embury. Dietz purchased Embury's tools and the Air Pilot name in 1953. The Dietz Air Pilot is not similar to any of the Embury lanterns. The Dietz Air Pilot is exactly the same as the streamlined D-Lite of Plate 8.10 but with a smaller No. 1 burner and smaller globe. The globe must be smaller to match the burning characteristics of the smaller wick. The Dietz Air Pilot uses a special 9 cm by 10.5 cm (3.5 in. x 4.25 in.) globe. Some Embury built Air Pilots are shown in Plates 8.4 and 8.5.

The U. S. built Dietz Air Pilots have "N.Y. U.S.A." embossed on the fount next to the fill. The design was revised slightly for production in Hong Kong starting around 1963.

This example is built using many of the original Dietz presses in China. It has the classic Dietz metallic blue paint. Imported by V&O of Syracuse, NY starting in 1992. The retail price in 1996 was $18.78.

This Air Pilot is the only Art Deco lantern in the line of eight imports. The other seven lanterns are: the No. 1 Little Wizard, No. 10 Monarch, No. 20 Junior, No. 50 Comet, No. 80 Blizzard, No. 90 D-Lite and The 1976 Original.

Variations:

Also available in solid brass and electrified.

Plate 8.14
Dietz Air Pilot, No. 8, 1953, 13.75 in.(34.9 cm) 7.75 in.(19.7 cm) $20-$30

Plate 8.14a
Label from a modern Dietz lantern.

Dietz Junior No. 20

Plate 8.15 **Description**:

This imported Junior is an all steel, cold blast lantern. It has a large, two piece fount cap and steel, No. 1, domed burner with the bent wire adjust knob and 0.625 inch (1.6 cm) wick. The Junior has an inside lift and a dome fount. The globe is not embossed (smooth) and has the Dietz logo painted on. Imported from Hong Kong and China.

Markings: (embossed on fount)

DIETZ No. 20
JUNIOR

(embossed on crown)

DIETZ (logo)

(painted on globe)

DIETZ (logo)

(gold label on bottom)

Made in the People's Republic of China

Patent Information: (none)

Remarks:

The Dietz Junior has remained almost unchanged for 100 years. The junior size globe was used by Dietz and many other lantern manufacturers. The Junior was marketed to women and even young children. Juniors came in dash, wagon, and inspector models just like the larger No. 0. The similarity between this modern Junior and the very first Junior is a tribute to the designer's skill so long ago.

This Junior has its original tin plating. It was first imported from China by V&O of Syracuse, N.Y. after R. E. Dietz Ltd. went out of business in 1992. The retail price was $15.60 in 1996.

All imported Dietz lanterns have a number embossed on the fount in the location where the American made lanterns say "N.Y. U.S.A." The imported Junior is designated No. 20 by Dietz. Before 1986, the Junior was made in the R. E. Dietz Ltd. factory in Hong Kong.

Variations:

The imported version of the Junior is found in tin plate, solid brass, antique bronze, red, and blue. In 1989, Dietz imported a solid brass collector's limited edition with individual serial numbers for their 150th anniversary (see ad in Plate 1.12).

Plate 8.15
Dietz Junior No. 20, 1956, 12.25 in.(31.1 cm) 5.25 in.(13.3 cm) $20-$30

Plate 8.15a
The No. 20 Junior, electrified and weathered, lights the shooting galleries at Disneyland.

Dietz Little Wizard No. 1

Plate 8.16 **Description**:

The all steel Little Wizard is a cold blast lantern with a large, two piece fount cap embossed with the Dietz logo. It has a little wizard size globe, inside lift, all steel, rising cone burner, with a 0.625 inch (1.6 cm) wick and bent wire wick adjust. The air tubes are machine embossed for strength. Machine crimp construction is used throughout. Imported from Hong Kong and China.

Markings: (embossed on fount)
DIETZ NO. 1 LITTLE WIZARD
(embossed on crown)
DIETZ (logo)
(painted on globe)
DIETZ (logo)
(gold label on bottom)
Made in the People's Republic of China

Patent Information: (none)

Remarks:

In production since 1920, the imported Dietz Little Wizard is made using the same dies as the Little Giant of Plate 7.32. The design goal for the Little Wizard was to make a small, long burning light for contractor's use. The Dietz Little Wizard has the largest fount available and uses a No. 1 burner. The No. 100 of Plate 8.11 is the longest burning with the same fount but an even smaller 0.38 inch (0.95 cm) wick.

In this example, imported from China, the globe is no longer embossed but is painted with the Dietz logo.

This Little Wizard has red paint and flashy brass trim. This is a far cry from the red globed and roughly handled Little Wizards of the past. It was imported from China by V&O of Syracuse, N. Y. after R. E. Dietz Ltd. went out of business in 1992. The retail price was $19.04 in 1996.

All imported Dietz lanterns have a number embossed on the fount in the location where the American made lanterns say "N.Y. U.S.A." The following numbers appear on the imported lanterns: No. 1 Little Wizard, No. 8 Air Pilot, No. 10 Monarch, No. 20 Junior, No. 50 Comet, No. 80 Blizzard, and the No. 90 D-Lite. The Dietz Original '76 has always been imported, first from Hong Kong and later from China.

Variations:

Available in solid brass, red, and the Dietz metallic blue paint.

Plate 8.16
Dietz Little Wizard No. 1, 1956, 12.0 in.(30.5 cm) 7.75 in.(19.7 cm) $20-$30

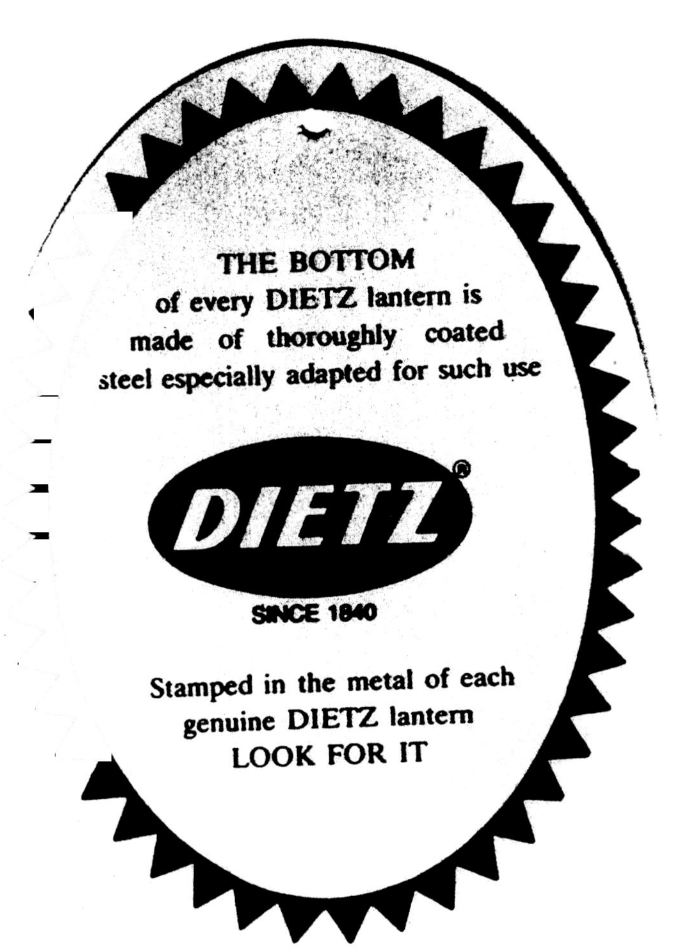

Plate 8.16a
Back of the red and yellow tag found on all modern Dietz imports.

Dietz Blizzard No. 80

Plate 8.17 **Description**:

This Blizzard is an all steel, cold blast lantern that has a large, two piece fount cap embossed with the Dietz logo. It has a No. 0 globe, inside lift, all steel No. 2, domed burner, with a 0.875 inch (2.2 cm) wick and bent wire wick adjust. The air tubes are machine embossed for strength. Machine crimp construction is used throughout. Imported from Hong Kong and China.

Markings: (embossed on fount)

DIETZ NO. 80
BLIZZARD

(embossed on crown)

DIETZ

(painted on globe)

DIETZ

(gold label on bottom)

Made in the People's Republic of China

Patent Information: (none)

Remarks:

The Blizzard was first produced in 1898, then redesigned in 1912 and 1939. This Dietz Blizzard was reintroduced as an import from the Dietz factory in Hong Kong. This Blizzard is essentially the same as the U.S. Blizzard of Plate 7.14. The globe is no longer embossed but is painted with the Dietz logo. This 1996 example was built without solder in China and is the largest lantern in the line. The Dietz Blizzard produces one of the brightest lights by using the largest globe and wick (0.875 inch) available. It is the tallest Dietz hand lantern.

All imported Dietz lanterns have a number embossed on the fount in the location where the American made lanterns say "N.Y. U.S.A."

The numbers on the imported lanterns do not appear to have any meaning. Collectors should continue to refer to lanterns by their name, not these numbers.

Imported by V&O of Syracuse, NY starting in 1992. The retail price was $18.18 in 1997.

Variations:

Available in Dietz metallic blue, red and bronze plated.

Plate 8.17
Dietz Blizzard , No. 80, 1956, 15.0 in.(38.1 cm) 7.75 in.(19.7 cm) $20-$30

Dietz D-Lite No. 90

Plate 8.18 **Description**:

This all steel, cold blast lantern has a large, two piece fount cap embossed with the Dietz logo. It has a short globe, inside lift, all steel No. 2, domed burner, with a 0.875 inch (2.2 cm) wick and bent wire wick adjust. The air tubes are machine embossed for strength. Machine crimp construction is used throughout. Imported from Hong Kong and China.

Markings: (embossed on fount)

DIETZ NO. 90 D-LITE

(embossed on crown)

DIETZ

(painted on globe)

DIETZ

(gold label on bottom)

Made in the People's Republic of China

Plate 8.18
Dietz D-Lite No. 90, 1956, 13.5 in.(34.3 cm) 7.75 in.(19.7 cm) $20-$30

Patent Information: (none)

Remarks:

Originally produced in 1913, the D-Lite was restyled in 1919 and again in 1939. This D-Lite then produced as an import from the Dietz factory in Hong Kong. The No. 90 D-Lite is essentially the same as the 1919 D-Lite of Plate 7.18.

To understand why Dietz reintroduced the 1919 D-Lite design we need to review the 1953 acquisition of the Embury Company. Embury, at the time, had three lanterns named Air Pilot; the No. 0 (Plate 8.6), the No. 2 (Plate 8.4) and the Little Air Pilot (Plate 8.5). Dietz wanted to continue to sell to customers who previously ordered Air Pilots. The answer was to modify existing tooling from the 1939 D-Lite and rename the result Air Pilot. This left a need for a lantern named D-Lite for customers who had been buying them for decades. The 1919 D-Lite had the same performance and the tooling had been retired a few years before.

This 1996 example, built entirely without solder in China, uses a short globe. The globe is no longer embossed but is painted with the Dietz logo. This lantern has the rising cone burner of the original D-Lite that was dropped in 1919.

This lantern has the classic Dietz metallic blue paint. Imported by V&O of Syracuse, NY starting in 1992. The retail price was $17.04 in 1997.

The numbers on the Chinese lanterns do not appear to have any meaning.

The name D-Lite is taken from a successful line of lanterns built by Dietz from 1913. The short globe was designed by Warren McArthur Jr. when he was the Dietz U.S. Sales Manager in the Chicago office.

Plate 8.18a
All imported Dietz lanterns have numbers embossed on the fount in the location where American made lanterns say "N.Y. U.S.A."

Dietz Monarch No. 10

Plate 8.19 **Description**:

The modern Dietz Monarch is available with and without brass trim. The cross guard version is a dead ringer for the 1913 hot blast lantern. The body is steel and it has a two piece fount cap and a steel, No. 1 burner with a 0.625 inch (1.6 cm) wick. The Monarch still has a rear globe lift, high dome fount top, and machine crimp construction. The air tubes are machine embossed for strength. Imported from Hong Kong and China.

Markings: (embossed on fount)

DIETZ No. 10 MONARCH

(embossed on crown)

DIETZ (in oval)

(painted on globe)

DIETZ (in oval)

(gold label on bottom)

Made in the People's Republic of China

Plate 8.19
Dietz Monarch , No. 10, 1956, 13.75 in.(34.9 cm) 5.25 in.(13.3 cm) $20-$30

Patent Information: (none)

Remarks:

The original Dietz Monarch was redesigned in 1913 and produced until 1950. This recreation was first produced in the Hong Kong factory purchased by Dietz Co. in 1956. Compare to the 1913 Monarch in Plate 7.19. The most obvious difference between the original and this reproduction is in the globe guard. On the original, the No. 0, LOC-NOB globe fits snugly into the globe guard wires. This reproduction has a smooth globe and a removable, triple wire guard. This Dietz Monarch had a retail price of $20.02 in 1996.

This example is painted black with contrasting brass trim. Imported by V&O of Syracuse, NY starting in 1992.

This lantern can be seen outside the temple of the "Indiana Jones" ride, and other places at Disneyland. With a dusty paint job, this lantern is a reasonable approximation of lanterns used before 1900. Hollywood prop companies should snap these up for use in period movies, and westerns. The author believes this is the only hot blast lantern still in production anywhere in the world.

Variations:

This lantern is available in solid brass, red, blue, black, and electrified.

Plate 8.19a
The red and blue Monarchs have this more conventional globe guard.

Dietz "The Original"

Plate 8.20 **Description**:
What makes this steel with brass trim, cold blast lantern original is its left hand lift! The Original also has a two piece, steel fount cap and a tiny, steel, rising cone burner with a 0.5 inch (1.27 cm) wick. It has a high dome fount and all machine crimp construction (no solder). The fill cap, globe and crown have the DIETZ logo. The air tubes are machine embossed for strength. It is imported from China.
Markings: (embossed on crown)
DIETZ (in oval)
(on fount)
DIETZ '76 THE ORIGINAL
(painted on globe)
DIETZ (in oval)
(impressed on bottom, 1976-1986)
MADE IN HONG KONG
(gold label on bottom, after 1986)
Made in the People's Republic of China
Remarks:
The Original '76 was introduced for the United States bicentennial in 1976. It is not a reproduction of an earlier Dietz lantern but a new design produced in the Dietz Hong Kong factory.

The left hand lift is very unusual yet it makes sense. A right handed person would hold the match in the right hand while lifting the globe with the left hand. The lift on this lantern is a pretty good indication that Dietz picked up the tooling from another company. It bears a striking resemblance to the older Feuerhand in Plate 9.5. The tubes are the same and the rising cone burner is identical. Did Dietz buy the dies from Feuerhand and use them to press the original?

The Original is smaller than a Junior (Plate 8.15) and larger than the Comet (Plate 8.9). Despite the scaled down dimensions, The Original gives a good light once it gets warmed up.
Variations:
This lantern was available in solid brass, red, and white.

Plate 8.20
Dietz The Original, 1976, 10.25 in.(26.0 cm)
4.5 in.(11.4 cm) $10-$20

Plate 8.20a
The Dietz "Original," second from left, shares many of the design features of the Chalwyn Tropic, left, the Feuerhand Baby, and the Winged Wheel No. 400, right.

Plate 8.20b
The "Original" comes in white, red, and solid brass.

Chapter 9

Foreign Lanterns

This chapter explores the details of foreign and Canadian kerosene lanterns for comparison to the U.S. designs. Foreign lanterns have been imported since the 1920s and, to a lesser extent, brought to this country by emigrants and antique dealers. A significant number of old European lanterns are found on the market today because many dealers specialize in buying container loads of antiques from Europe and selling them in the U.S.

There are no kerosene lantern manufacturers in the U. S. today, yet new lanterns abound. The major suppliers of new lanterns are China, Hong Kong, and Taiwan. Other current lantern making countries include Kenya, France, and India. Japan was a big importer in the 1960s and 70s which explains why Winged Wheel brand lanterns are so plentiful but, the balance of trade shifted and I expect they now import from China too.

Foreign lanterns add interest and variety to any collection.

Table 9.1, Foreign Lanterns

PLATE	LANTERN MAKER/NAME	DATE	TYPE /STYLE	COUNTRY	OVER-ALL HEIGHT	OVER-ALL WIDTH	GLOBE TYPE	WICK(in.)
9.1	Sherwoods Ltd.	ca.1912	hotblast	Britain	14.0 in.(35.6 cm)	6.0 in.(15.2 cm)	British	.75"
9.2	Trulite	ca.1920	coldblast	Canada	14.25 in.(36.2 cm)	6.5 in.(16.5 cm)	short	.875
9.3	Feuerhand Nier	ca.1930	coldblast	Germany	13.5 in.(34.3 cm)	7.0 in.(17.8 cm)	special	1.0"
9.4	Angemeldet 682	ca.1930	coldblast	Germany	10.25 in.(26.0 cm)	5.25 in.(13.3 cm)	special	0.5"
9.5	Feuerhand Baby	ca.1950	coldblast	Germany	10.0 in.(25.4 cm)	6.0 in.(15.2 cm)	original	0.5"
9.6	Chalwyn Tropic	ca. 1955	coldblast	Britain	9.5 in.(24.1 cm)	6.0 in.(15.2 cm)	original	0.5"
9.7	Winged Wheel No. 400	ca. 1960	coldblast	Japan	9.5 in.(24.1 cm)	6.0 in.(15.2 cm)	original	0.5"
9.8	Kwang Hwa 245	ca.1980	coldblast	China	7.5 in.(19.1 cm)	4.8 in.(12.2 cm)	special	.25"

Unmarked imports like this are available in hardware stores around the world.

Sherwoods Limited

Plate 9.1 **Description**:

This British hot blast lantern is mostly steel. Many details differ from American designs like the fount cap is internal thread and the wick is 0.75 inch (1.9 cm) wide. The globe shape differs from its U.S. counterpart but the top lift is similar to early American designs. Sherwoods used machine crimp construction with some soldered parts. The air tubes are seamless tubing. It has no wire globe guard and the globe plate is hinged to allow access to the burner. Made by Sherwoods Limited in Birmingham, England.

Markings: (on crown)

SHERWOODS BIRMINGHAM
BRITISH MADE
(on globe)
BRITISH MADE
(on knob)
SHERWOODS LTD. BIRMINGHAM

Plate 9.1
Sherwoods Ltd., ca. 1912, 14.0 in.(35.6 cm) 6.0 in.(15.2 cm) $20-$30

Remarks:

The Sherwoods lantern has no model number or name to differentiate it from the other lanterns made by the company. Perhaps this was the only lantern model they made.

Notice the globe shape. The size of the globe is comparable in size to a U.S. No. 0 but the shape is quite different. The air tubes are plain tubing. The correct finish for this lantern is unknown.

If this lantern ever had a globe guard it would have been a removable type much like the Chinese built Monarch of Plate 8.19.

The 0.75 inch wide wick is unusual here in America but more that, it is 10 inches (25.4 cm) long!

Trulite

Plate 9.2 **Description**:

This Canadian lantern similar to U.S. made lanterns of the same period. It is an all steel, cold blast lantern, with a large, one piece fount cap and brass, No. 2, domed burner with a 0.875 inch (2.2 cm) wick. The Trulite has a short type, ruby globe and an inside lift.

Plate 9.2
Trulite, ca. 1920, 14.25 in.(36.2 cm) 6.5 in.(16.5 cm) $20-$30

Markings: (embossed on bell)
Trulite
(embossed on fill cap)
Made in
HAMALITON
CANADA
(nothing on the fount)
Patent Information: (none)
Remarks:

It is clear, from this lantern at least, that Canadian construction matches U.S. practices. The Trulite uses a standard "short" globe invented by Warren McArthur Jr. in 1912. The No. 2 burner is interchangeable with U.S. models. The bail and the large, one part fill cap are unusual but everything else is conventional. The air tubes have the same design as the 1913 D-Lite. The overall styling looks top heavy because of the small fount.

Plate 9.2a
Trulite script logo.

Feuerhand Nier

Plate 9.3 **Description**:

The Feuerhand Nier is a German-made, all steel, cold blast lantern for Kerosene. It has a steel burner with a 2.5 cm (1 full inch) wick. The Nier has an extra large, 3 cm, fill cap with restraining wire, tin finish, and all crimp construction. The globe dimensions are identical to the "short" globe of Table 4.2. Using a standard American globe is a good idea in an imported lantern. Made in pre-war Germany.

Markings: (embossed on crown)
FEUERHAND *
(embossed on bell)
NIER
FEUERHAND (logo) FIREHAND
REG. U.S. PAT. OFFICE
(embossed on fount)
MADE IN GERMANY
FEUERHAND (logo) * Nr. 260 *
(embossed on globe, burner adjust and cap)
(logo)
FEUERHAND MADE IN GERMANY

Remarks:

The Feuerhand Nier was made for import to the U.S. as it has markings in English and a U.S. patent warning. Nier means "near" and Feuerhand means "Firehand" which matches the logo.

The 1 inch wick is unusual in the U.S. and probably a bad idea in an imported lantern. Standard American wicks are 0.625, 0.875 and 1.5 inches.

This lantern would have been made after the recovery from World War 1 and before the trade embargo of World War 2. After W. W. 2 this lantern would have been marked "West Germany."

The wire restraint on the fount cap is common on foreign lanterns and rare on U.S. models. R. E. Dietz only did a few restrained caps between 1912 and 1936. Perhaps the owner of the patent wanted too much for the rights.

Plate 9.3
Feuerhand Nier, ca., 1930, 13.5 in.(34.3 cm)
7.0 in.(17.8 cm) $20-$30

Plate 9.3a
Detail of the Nier's bell.

Angemeldet No. 682

Plate 9.4 **Description**:

The Angemeldet No. 682 is a small, all steel, cold blast lantern. It has a small, one piece fount cap and steel, domeless burner with a 0.5 inch (1.27 cm) wick. Pulling down on the bale lifts the chimney off the globe so the spring mounted globe plate can be tipped off the burner. The special size small globe is clear. Judging by the English markings, the No. 682 was made for export, in Germany, before 1945.

Markings: (embossed on fount)
Angemeldet AUTOMATIC -
ASA 682 DRP
MADE IN GERMANY ASA No 682

Patent Information: (none)

Remarks:

The Angemeldet No. 682 is a small lantern suitable for use inside your chalet.

When the bale is lowered it engages loops on the chimney to lift it. The globe plate sits on a spring that must be compressed as the globe is tilted back off the domeless burner. This must be done each time the lantern is lit as well as for burner maintenance. The fill hole is unusually deep compared to domestic examples. The Angemeldet is an odd size that is larger than a Dietz Original and smaller than a Junior.

The features of this lantern place it between W. W. 1 and the end of World War 2, This lantern may have witnessed the rise and fall of Nazi Germany.

The original finish appears to be a gloss black enamel. It is not clear what ASA stands for.

Plate 9.4
Angemeldet 682, ca. 1930, 10.25 in.(26.0 cm) 5.25 in.(13.3 cm) $20-$30

Plate 9.4a
Angemeldet logo.

Feuerhand Baby

Plate 9.5 **Description**:

The Feuerhand Baby is an all steel, cold blast lantern for kerosene. The inside lift is on the left side. It has a steel burner with a 0.5 inch (1.3 cm) wick. The Baby has a two part fill cap embossed with the Firehand logo. The finish is tin and it has all crimp construction (no solder). The globe is similar to the "original" globe of Table 4.2. The Baby was made in West Germany for export.

Markings: (embossed on crown)
* ORIGINAL - NIER - FEUERHAND *
(painted on globe)
FEUER (logo) HAND
MADE IN W. GERMANY
Nr. 1275 ® Nr. 1276
JENA ER GLAS
SCHOTT
MAINZ
JENA ER GLAS
(embossed on fount)
WESTERN GERMANY
*275
BABY
FEUER (logo) HAND
(embossed on fill cap)
(Feuerhand logo)

Patent Information: (none)

Remarks:

The Feuerhand No. 275 BABY was made for import to English speaking countries as it has markings in English and German. Nier means "near" and Feuerhand means "Firehand."

Besides the left side lift, the bent wire wick adjust shape is also unusual. The Baby has "ears" for the bail as do most European lantern designs.

This lantern would have been made after the recovery from World War 2 by a company that was making lanterns before the war.

Many foreign lanterns have a chain or wire restraint on the fill cap so it does not get lost. The BABY does not have any restraint. Most foreign lanterns have the bail attached to the air tubes by "ears" which was also not done by U.S. companies.

Plate 9.5
Feuerhand Baby, ca. 1950, 10.0 in.(25.4 cm)
6.0 in.(15.2 cm) $20-$30

Plate 9.5a
The descriptive Feuerhand (Firehand) logo.

Chalwyn Tropic

Plate 9.6 **Description**:

The Chalwyn Tropic is an all steel, cold blast lantern, with a 0.875 inch, two piece, unrestrained fount cap embossed with CHALWYN MADE IN ENGLAND. It has a small globe, right hand inside lift, domed burner, with a 0.5 inch (1.3 cm) wick and bent wire wick adjust. The air tubes are machine embossed for strength. Machine crimp construction is used throughout.

Markings: (embossed on fount)
CHALWYN
MADE IN ENGLAND
TROPIC
(embossed on crown)
TROPIC
MK2
JA12159
1954

Patent Information: (none)

Remarks:

This product of Britain has formal lines and a hunter green coat of paint over the tinned steel. The shape of the globe matches the shape of the air tubes.

The Tropic may not be an import to America as a dealer brought it over with a load of antiques. The Tropic produces a bright and even light with its Dietz "original" size globe and 0.5 inch (1.3 cm) wick. It is about the same size as the Dietz Original, the Feuerhand BABY, and the Winged Wheel No. 400.

Does the 1954 on the crown refer to the first production year? Maybe. The design and condition are consistent with the 1950s through 1970s.

Compare the Tropic to the similar size Dietz Original of Plate 8.20.

Plate 9.6
Chalwyn Tropic, ca. 1955, 9.5 in.(24.1 cm) 6.0 in.(15.2 cm) $20-$30

Winged Wheel No. 400

Plate 9.7 **Description**:

The Winged Wheel No. 400 is an all steel, cold blast lantern, with a 0.75 inch (1.9 cm), two piece, unrestrained fount cap. It has an "original" size globe, inside lift, all steel, domed burner, with a 0.5 inch (1.3 cm) wick and bent wire wick adjust covered by a stamped tin cap. The air tubes are machine embossed for strength. Machine crimp construction is used throughout. Imported from Japan.

Markings: (embossed on fount)
Winged Wheel No 400
MADE IN JAPAN
(Winged Wheel logo)
(embossed on wick adjust and fill cap)
(Winged Wheel logo)

Patent Information: (none)

Remarks:

This product of Japan is one of several imported lanterns from Winged Wheel. The No. 350 is the size of a Dietz Comet, the No. 400 is the size of the Dietz Original, and the No. 2000 is the size of a Junior lantern.

The Winged Wheel lanterns are usually painted red but other colors, yellow for one, are seen from time to time. These lanterns were imported in large numbers in the 1960s and 1970s and gave R. E. Dietz Company a run for the money.

Compare to the Dietz Comet of Plate 8.9, the Dietz Junior of Plate 7.25 and the Dietz Original of Plate 8.20.

Plate 9.7
Winged Wheel No. 400, ca. 1960,
9.5 in.(24.1 cm) 6.0 in.(15.2 cm) $10-$20

Plate 9.7a
Winged Wheel logo.

Kwang Hwa

Plate 9.8 **Description**:

The Kwang Hwa 245 is an all steel, cold blast lantern, with a large, two piece, chain restrained fount cap embossed with Chinese characters. It has a tiny globe, inside lift, all steel, domed burner, with a 0.25 inch (2.2 cm) wick and bent wire wick adjust. The air tubes are machine embossed for strength. Machine crimp construction is used throughout. Imported from China.

Markings: (embossed on fount)
245
(additional Chinese characters)
(embossed on crown)
"KWANG HWA" MADE IN CHINA
(additional Chinese characters)
(embossed on globe)
(Chinese characters)

Patent Information: (none)

Remarks:

This product of China suggests it was imported after the détente of the 1970s. This lantern is the smallest functional lanterns in this book. Once warmed up, this lantern produces a bright and even light with it's tiny 0.25 inch (0.64 cm) wick. The globe must be smaller to match the burning characteristics of the smaller wick.

This example has the remains of the original tin finish. Lanterns like these are currently available in a variety of painted colors from gift shops and catalogues. It is hard to tell the size from a photo like this. Always check the size carefully when ordering from a catalogue to get a clear understanding of the lantern's actual size.

Compare to the smallest U.S. lantern, the Dietz Comet of Plate 8.9 and largest hand lantern, Defiance No. 200 of Plate 7.29.

Plate 9.8
Kwang Hwa 245, ca. 1980, 7.5 in.(19.1 cm)
4.8 in.(12.2 cm) $10-$20

Chapter 10

Repair and Restoration

Should I Restore?

I think we have a responsibility to care for antiques while they are in our custody. Generations before us took care of them and generations after us will do the same. Our responsibility is to maintain the condition during our time on earth. I would never suggest repainting a Ming vase or a Van Gogh but lanterns are more like old cars than fine china. Cars are made of steel that rusts if not protected. If the car is cut up or rusts away, it's gone forever. It does not matter so much what color the car is painted because in 40 or 60 years it will need repainting anyway. The same is true for lanterns. Never cut the metal or allow a lantern to rust as this destroys the historic value. Go ahead and paint the lantern any color you want because paint protects the lantern from rust and paint is easily stripped and redone by the lantern's future owners. However, there are some things to consider before grabbing a wire brush.

Experts agree that examples in fine original condition should be preserved as is, and not restored. The value of a rare lantern in exceptional original condition can actually be reduced by even the most expert restoration. A 150 year old lantern looses much of its appeal with a modern coat of paint. The original finish is obscured and the true condition, covered up. It would be like putting a new coat of varnish on a Stradivarius violin. Replace the missing cap or burner, sure, but leave the weathered metal to tell its story. Good lanterns that will be stored safely out of the weather, or displayed indoors, do not require restoration as their deterioration is temporally halted.

Decorators recommend that lanterns with an interesting patina should not be restored. This would include functional lanterns with multiple layers of cracked and peeling paint and a little rust here and there. These lanterns find their way into restaurants, boutiques, and homes as object d' art.

Do not despair. There are plenty of lanterns that have no sign of the original finish and are just plain rusty. Some rare lanterns are in such poor condition that restoration is necessary to prevent their total loss. These lanterns are ripe for preservation. Just remember to consider the interest, beauty, and value of maintaining the original finish before charging into a restoration.

Restoration

Lanterns were designed and built for hard use, but the ravages of time have left most early examples in poor condition. A rusted fount is the most difficult problem to resolve. A few small holes can be soldered closed for a classic repair. Unfortunately when a lantern has one or two weep holes in the bottom, it is a sure sign that more holes are on the way. Rust on the outside of the bottom is the result of allowing a lantern to sit on the ground trapping moisture underneath. Rust inside the fount indicates the lantern was empty and uncovered, allowing moisture to collect inside. A lantern hung on a nail in a drafty attic will age very gracefully.

Even a lantern partially rusted away can be salvaged. The Hurwood Aladdin of Plate 7.8 was just such a phoenix. Since no paint remains on this lantern it is a sure bet that the original finish was natural tin. If the lantern had ever been painted, a small bit of paint would remain in some out of the way spot. Color inside the burner will often give away the original finish because it has more protection from the weather. The Hurwood required the following process to preserve it for the future.

Rust and Paint Stripping

Chemical stripping, wire brushing, and bead blasting are all tools that can be used for a first class restoration. Each of these tools solve the same problem in a different ways.

Wire brushing is done with a wire wheel mounted to a high speed motor. There is some risk to the operator and to the lantern but the tools are inexpensive. The results can be quite good if you're willing to follow up with sandpaper in the corners where the brush cannot reach. Motors, wire wheels and accessories are available from the suppliers listed in this book.

Sand blasting should be avoided because antique sheet metal parts can be stretched, hardened, and distorted by careless blasting using silicone carbide or aluminum oxide. Bead blasting uses glass bead as the abrasive which is best for soft metals like brass, aluminum, and the tinned steel used in lanterns. Good results require an experienced operator and the equipment is expensive. Bead blasting equipment, media and accessories are available from the suppliers listed in Appendix B.

Chemical stripping is a gentler process that saves the surface finish of the metal and only removes paint and rust. I heard that chemical stripping was expensive, but compared to the mess of paint strippers, wire brushes, sandpaper, my time, dirt, grease, and dust, I found professional stripping is a bargain. Take a pile of dirty, greasy, rusty lanterns down to the stripper one day and pick up clean, bright metal lanterns a week later. The cost may be higher than wire brushing but it is the only method that removes rust inside the fount and tubes.

There are some drawbacks to chemical stripping that makes sandblasting better than chemicals for some parts. Aluminum, though not usually found on lanterns, is instantly dissolved in the Caustic Soda dip. Another problem is the caustic solution is a powerful base that is difficult to completely remove. The Caustic Soda continues to attack the metal if not properly neutralized by your stripper and again by you. The caustic solution gets everywhere, under rivets and inside seams. If the chemicals that worked their way into the seams while the lantern soaking in the tank, are not completely neutralized, the metal will continue to react and corrode causing damage the metal and/or the finish.

To prevent these problems you must not depend on the strippers neutralizing bath but clean each seam with a mild acid. I use a metal etch that contains phosphoric acid to neutralize the caustic base and to prepare the metal for primer. The phosphoric acid can be rinsed off with water and neutralized with a baking soda solution. Remember the rust stripping process is to work from very strong bases to very mild acids, like water, and always wear eye and skin protection. With the metal clean it is always a race to get the primer on before new rust begins.

Plate 10.1
There are several ways to remove rust including pickling in a mild acid.

Patching

The most common lantern repair is the dreaded leaky fount. There is no "best" cure but a simple solution for this tricky problem is automotive gas tank sealer. This sealer is designed to stick to clean metal, fill small holes, and resist hydrocarbons like gasoline and kerosene. Gas tank sealer is available from antique car restoration suppliers. Just follow the directions and don't forget to clean the sealer from the fill cap threads before it hardens.

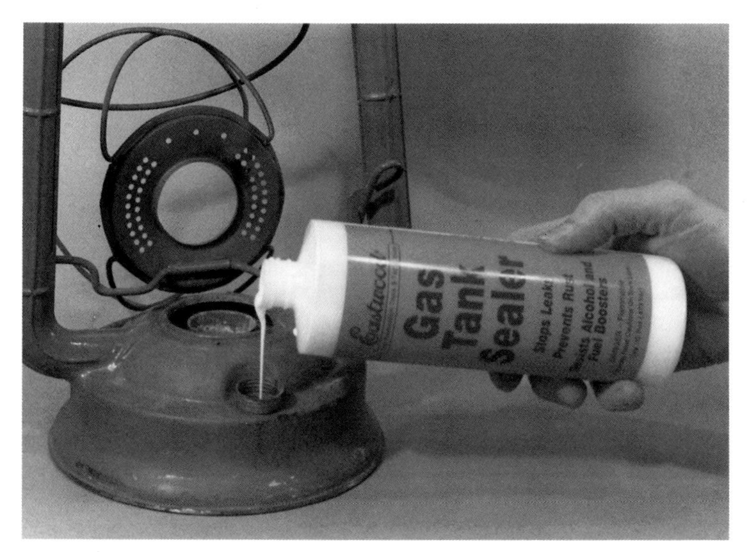

Plate 10.2
Gas tank sealer is a quick and easy way to seal a leaky fount.

Holes in the fount can be patched by a more authentic method if you are handy with a soldering iron. Fill the fount with water and mark each leak. Dimple each hole slightly with a dull punch to make a depression for the solder to fill. An acid flux must be used to get solder to stick to steel. Acid flux and solder are available from restoration suppliers or from the plumbing department of your hardware store. Make it easy on yourself and buy a water soluble flux. A 50 Watt or larger soldering iron is recommended. Plate 10.3 shows a solder repair on a removable signal lantern fount. A small torch can be used but in tight places a torch sometimes unsolders more than it solders. Apply the flux and then fill each dimple with a skin of solder. Use a file to smooth out bumps for a permanent and almost undetectable repair.

If there are too many holes for solder alone then a patch of brass sheet can be soldered in place. Brass sheet is available in craft and hobby stores.

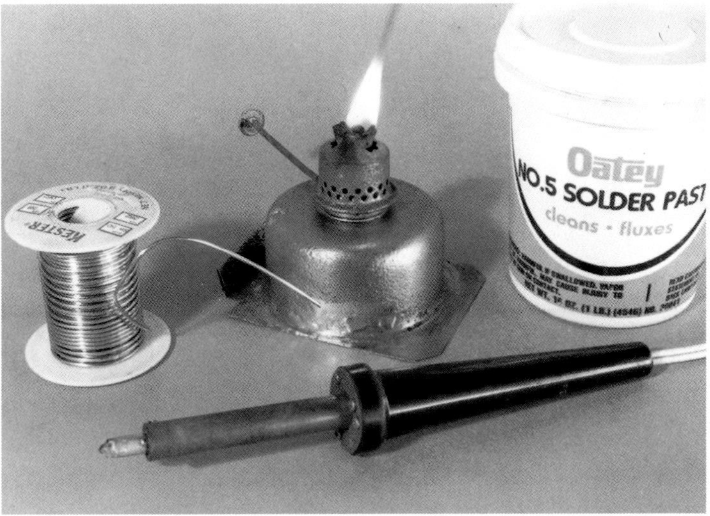

Plate 10.3
Flux paste and a large soldering iron makes quick work of a small hole.

Sometimes it is necessary to replace the entire bottom with a new disk of metal. The top or bottom of a one gallon paint can is an ideal source for this disk. To do this, remove the lantern bottom with tin snips but leave a half inch (1.0 cm) lip all around. Cut the paint can bottom so it fills the entire bottom of the lantern. Clean the contact surfaces of both pieces and add flux. If both parts are clean it is easier than it sounds to make a fuel tight seal.

Look for holes at the bottom of the draft tubes. Dirt and moisture tends to collect there and cause rust. Holes in draft tubes reduce the effective size of the flame before it begins to smoke. Filling the holes allows a brighter light and better flame appearance.

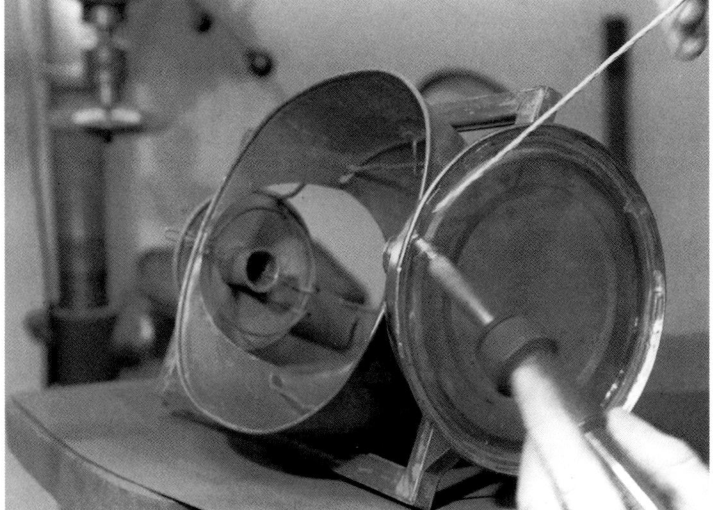

Plate 10.4
Replacing the entire bottom is sometimes required and not as difficult as it sounds.

An alternative to solder is epoxy. Available in craft and hardware stores this two part adhesive is strong and kerosene proof. Avoid "Fast" epoxy as they sometimes remain flexible after they cure. The metal needs to be clean or the adhesive will not stick. Mix it according to the directions on the label and fill the crack or hole. Once the epoxy has hardened the repair can be shaped with files and sandpaper.

It is always preferable to remove a dent rather than fill it. It is relatively easy to work dents out when there is access to both sides. Some dents can be pressed out with fingers or the handle of a screwdriver. Other dents require the more persuasive hammer and dolly approach (Plate 10.5). Unfortunately, in most lanterns we find the impossible-to-remove fount and tube dents. These dents are usually best left alone or filled.

A filler that may be useful for lantern repair is Bondo® Automotive Body Filler. Bondo® is a two part, polyester based resin combined with various fillers. Available in auto part stores, Bondo® is easy to mix, shape and finish but is somewhat porous and will not seal founts. Bondo® is useful for filling dents and depressions.

Filling dents raises an ethical question. Even if you have no plans to ever sell this lantern it may be sold someday. Filling a dent with solder, epoxy or Bondo® is unethical for resale. I sometimes prefer to leave them (see the Little Air Pilot of Plate 8.5) as the dents tell a story of the lanterns rough history. I believe that repairs required to make a lantern function are reasonable. No one can tell you what to do with your lantern. You must use your own judgment.

Plate 10.5
Special dent removal hammers and dollies are shown with sand paper and a sanding block.

Painting Philosophy

Few topics have as much mystery surrounding them as paint application. It's as close to an artistic process as any in lantern restoration. Some people have a natural talent and cannot explain their skill to others. The rest of us have to rely on sheer determination or settle for mediocre results. My approach is to expect flaws but not accept flaws. I am willing to sand and repaint the same lantern over and over until I am satisfied that it is as good as new. I found the hardest part of my approach is knowing when the result is good enough. On more than one occasion I sprayed a second coat on a lantern that was almost perfect, making it worse. I want to emphasize that my methods are not the easiest, fastest, or least expensive. Feel free to learn from my experiences, both failures and successes.

Priming

Self etching primer is very expensive. Professional paints, primers and thinners range in price from $35 to $85 per gallon. I decided not to mix different brands of paint but instead pick a "system" and stick with it. All major labels have the full spectrum of compatible products from metal prep to clear coat. I chose DuPont but I'm sure the other name brands are just as good. The important thing is to use compatible paints, solvents, and primers.

There are many good books on painting technique in the library. I must confess that I have read many books and received training from a professional refinisher but it didn't do a heck of a lot of good until I painted a bunch of lanterns. There is no substitute for practice and you can't practice on a piece of cardboard. Get yourself a few Dietz Monarchs and practice on them.

The restoration drill for all painted steel parts is as follows: After chemical strip or bead blasting, the part is straightened as close to the original shape as possible. Now is the time to make any solder repairs or patches needed.

In the days or weeks it might take to get the lantern ready to paint, the metal surface may be covered with a light coat of rust. Clean the oxidation from the surface with a metal prep (dilute

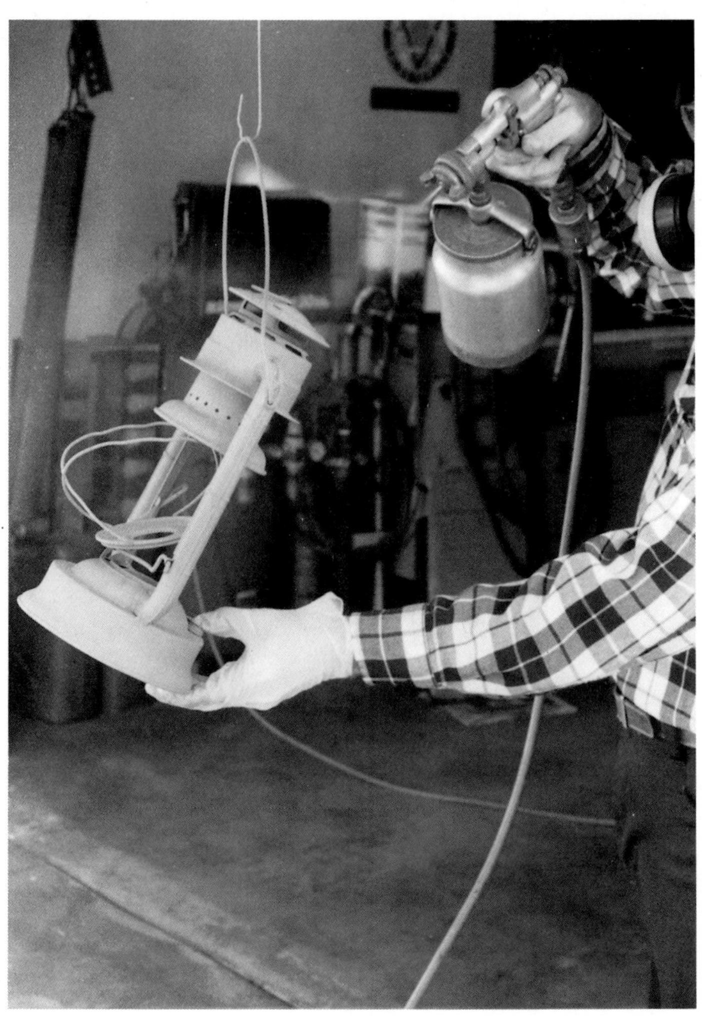

Plate 10.6
A good aerosol primer can work just as well as more professional primers.

phosphoric acid) and then neutralize the acid with a Baking Soda solution in water. Give the lantern a good rinse and dry with a lint free towel.

Removing the rust, dirt, grease, and grime is the most important step in achieving a durable and good looking finish. Carefully clean the lantern, in a well ventilated location, using a paper towel and a quick drying solvent like white gas, lacquer thinner, or enamel reducer. Now is the time to mask off any polished parts that should not be painted. The bail and the burner are usually not painted because the bail would quickly chip from handling and paint on the burner would burn off even quicker. Wipe away any lint, dust, or paper fibers with a tack rag. Tack rags are sticky cheese cloth available in any paint store.

Now comes the most important layer of paint, the primer. Follow the directions on the label. If the primer sticks to the metal, the paint will stick too. A top quality respirator, eye protection, and disposable gloves are always required when working with any kind of paint. Plate 10.6 shows a self etching primer being applied to a Dietz D-Lite of Plate 7.18.

If the lantern has major rust pitting then a paint called "surfacer" is required. Surfacer is just paint with a lot of filler to cover surface imperfections. Begin with one heavy coat of surfacer and then sand with 100 grit paper until the primer coat just begins to show through on the high spots. If there is still plenty of overall pits, another coat of surfacer is required. If there are a few places where the sand paper did not cut down to the low spots, a lacquer spot putty can be applied.

Some heavily pitted areas may be completely covered with putty at this point. To save on the sanding keep the putty as thin as possible. Spot putty is removed with 320 "wet and dry" sandpaper using lots of water and being careful not to remove the primer coat. A final sanding with 600 grit "wet and dry" gets the surface ready for the color coat.

The minimum required paint supplies are is as follows:

- Metal prep
- Tack Rag
- Primer
- Red Oxide Primer Surfacer
- Enamel or lacquer color paint
- cleaning solvent (Lacquer Thinner, etc.)
- 120 grit, 320 grit, and 600 grit sandpaper

Painting

With the first class preparation complete just about any paint will give excellent results. Automotive paints are preferred for their all weather endurance. Auto paints can also be mixed in any desired color. As a practical matter aerosol spray cans are the most convenient approach. Automotive supply outlets have a selection of colors, including metallics, that reproduce the original lantern colors very well.

Paint in spray cans is always very loose, thin and watery. The correct technique for spray painting is to start the spray off the target, then sweep across and stop the spray off the target. Stopping or starting on the lantern will make a heavy spot that will likely run. Because the paint is thin, it is always best to apply several light coats rather than a heavy coat.

Quickly spray two light coats using the instructions on the aerosol can. The tried-and-true process for spray painting a lantern is to spray the bottom surfaces first. First spray two light coats inside the crown, inside the chimney, the tube bottoms, and the fount bottom. Wait 15 minutes between coats of paint then, starting at the top, spray two light coats over the whole lantern. Always work from the inside to the outside and from the top down.

Because of the complex shape of a lantern there is always a danger that some areas will get multiple over spray as other areas are painted. Try to avoid this by keeping each coat as light as possible. The final coat should be all over, from top to bottom, and just a little heavier. It can be heavier because the paint wants to stick to itself. This is the beauty coat that gives the lantern a smooth, shiny finish. Allow the paint to dry in a dust free area.

Use the same spray technique to spray a clear coat over all polished parts. If a clear lacquer is not used, the brass and copper parts will quickly tarnish.

Plate 10.7
Automotive touch up paints come in a variety of colors and are very durable.

Plate 10.8
Simple home plating can be done using this electroplating kit.

Plating

Numerous lanterns described in this book originally had a bright tin, zinc, or galvanize coating to protect the steel from rust. Some of the lanterns are solid brass and others are brass with a protective nickel plating. None of these finishes can be effectively simulated with paint. Paint provides the rust protection but doesn't duplicate the beauty of new tin or polished metal.

Lanterns can be replated with some cost and difficulty. Only valuable lanterns in excellent condition are candidates for plating because of the expense. The cost of plating is relatively high when compared to a three dollar can of spray paint. The results achieved with paint does not compare to polished nickel or brass but, on the other hand, no permanent damage has been done. The good news is a brass lantern can be professionally polished or polished at home using inexpensive materials.

Be aware that buffing is necessary before the lantern is plated. The finish will be no better than the quality of the buffing.

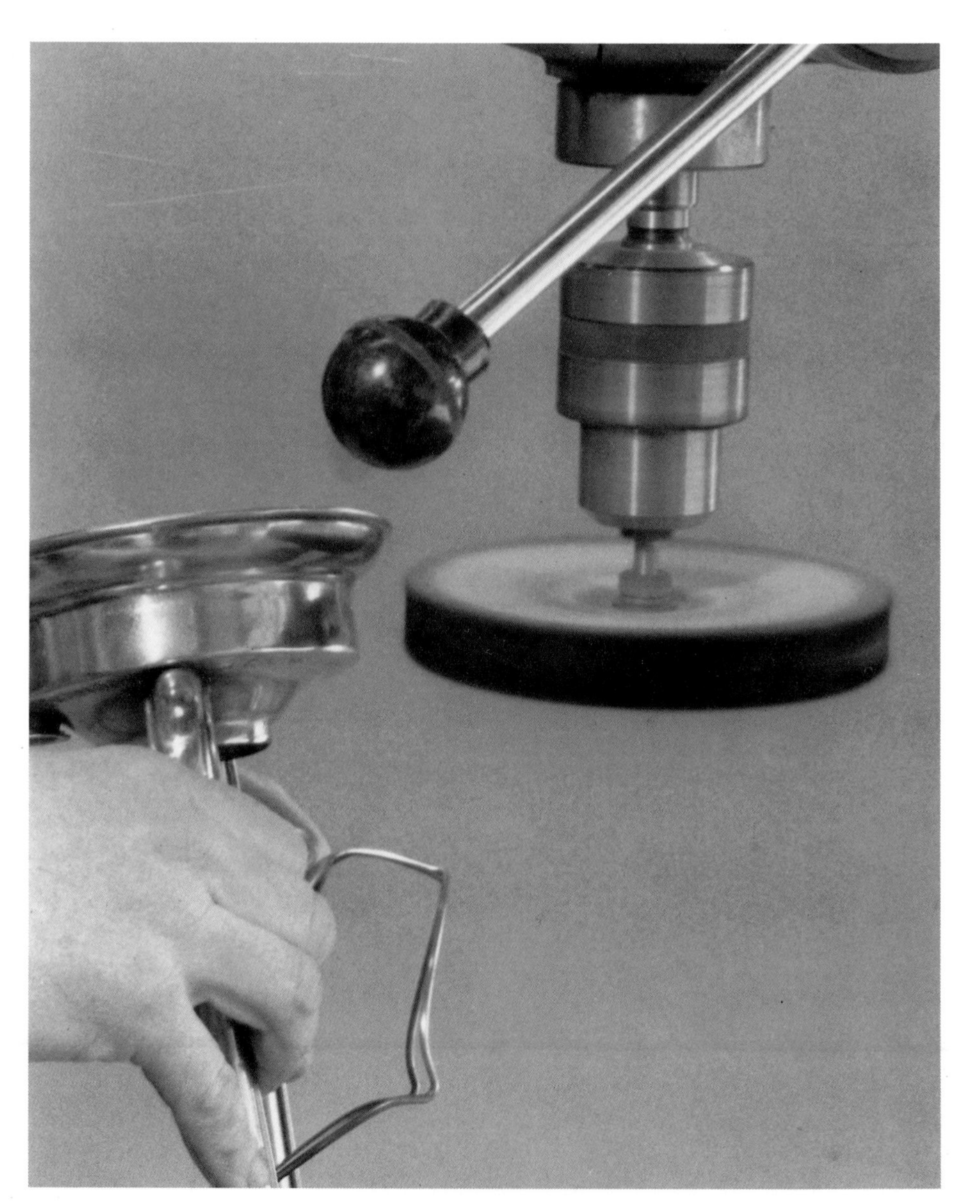

Plate 10.9
A loose section buffing wheel mounted in a drill press can give decent results.

Buffing

Some of the most beautiful and interesting lanterns described in this book are buffed solid brass. Many other lanterns had some brass or copper parts that should be buffed. The Embury Camlox of Plate 7.14 has a very unusual copper fount and the Buhl has a brass fount (Plate 7.23). If you are luckily enough to have a lantern with brass or copper parts, they likely need buffing.

If you are going to have a lantern professionally buffed then some knowledge of the process and hazards are in order. If you plan to tackle the buffing process yourself, read on.

Clarification of some terms are in order. What we know as polishing is actually buffing. When we polish the family silver we are buffing the silver. The factory polished the silver to make it smooth and then buffed it to make it shine. Professionals use the terms "polishing" and "buffing" in this way. Polishing takes metal off to give a smooth surface. Buffing cleans the surface for a bright appearance. It's important when talking to a professional to agree on the quality of finish desired.

Most kerosene lanterns were made of steel and either painted or finished with tin to prevent rust. Steel lanterns should be painted or tin plated and that's all. Some special railroad lanterns were plated in silver and even gold but it's doubtful that any barn lanterns were sold with such expensive finishes. Brass lanterns were sometimes nickel plated and these lanterns really shine (pun intended) when they are buffed. If a lantern is copper or brass then more possibilities present themselves.

A word of caution is needed here. There is a potential for damage to the lantern during any or all of these operations. Even professional platers and polishers make mistakes that can destroy your lantern. If the brass, copper, or nickel is in reasonably good condition, then it may be prudent to just clean and buff it by hand.

Most of the materials and tools needed to repair and restore lanterns are available from any well stocked automotive restoration supply source. One mail order source, Eastwood, has all the tools, paints, polishing wheels, rouge, and other supplies need to perform all the restorations described in this book.

Plate 10.10
Tripoli and white rouge are the most useful buffing compounds for soft metals like brass, copper, and zinc.

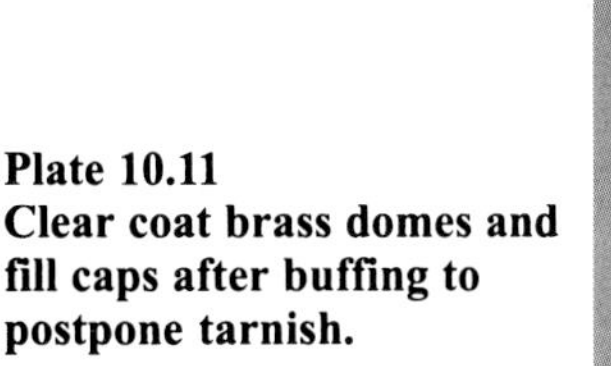

Plate 10.11
Clear coat brass domes and fill caps after buffing to postpone tarnish.

Chapter 11

Lantern Prices

When searching for lanterns only two things are certain: Old lanterns cost more where you are, and someone just bought all the good ones last week.

Determining a lantern's value is a very difficult task. The only correct answer is: a lantern is worth what someone will pay for it. Unfortunately, the appraiser needs a little more to go on before the lantern is put on the auction block or antique store shelf. People depend on price guides for accurate information in a hurry. Accurate historical information about kerosene lanterns has become widely available in recent years. A few years ago, most lantern prices were set strictly on an uneducated guess. I believe there still is wide spread ignorance about lanterns but, at least now, there are some excellent reference books for anyone who bothers to check. Several information and value guides are listed in Chapter 12.

There are many price guides for collectible kerosene lanterns. The thick annual guides list the results of sales and auctions from around the country but they lack a complete description of the lantern make, model and condition. The good thing about auction results is they represent what someone was willing to pay. The disadvantage is, lanterns are rarely sold at auction so there are not enough listings for an accurate appraisal.

Other price guides give a broad range of value based solely on the author's experience or opinion about what she/he thinks each lantern should be worth. The problem with some expert's opinion is it can be colored with the expert's bias like, how long they spent looking for a particular model or how many of a particular lantern was made. Quantity is a factor in value but not as important as demand.

In this chapter, we look at the specific factors that determine a lantern's worth and attempt to base the value on those specifics. The goal is a logical and comprehensive determination of any lantern's value.

Plate 11.1
It is educational to see the wide range of lantern prices at a railroadiana swap meet like this one at the Orange Empire Railway Museum in Perris, California.

Lantern Value Calculation

The Lantern Value Calculator (LVC) aids an unbiased appraisal first, of the condition, and then, of the lantern's relative worth. Two lanterns, side-by-side, can be now evaluated on their relative value. Once the condition is established, the ranking of its good and bad features can begin. Placing an actual value to the final score still requires the appraiser's skill.

Desirability

Many factors enter into the value of a collectible but DESIRABILITY and CONDITION are paramount. An antique in new condition is worthless if no one wants it. Likewise, a desirable antique in dreadful condition will never have much value.

Many assumptions are made in the development of all price guides and this is no exception. The difference is that I explain how the ratings are determined. Some of the assumptions used to develop this price guide are: older is worth more than newer, popular is worth more than undesired, rare is worth more than common, railroad styles are worth more than barn styles, unusual lanterns are worth more than plain designs, and so forth.

Some factors that cannot be included in this price guide are:

Railroad and contractor lantern prices vary according to the names of railroads and cities (respectively) embossed into the metal. Prices vary greatly by regional demand. For example, New York Central marks are more popular on the east coast and Southern Pacific prices are higher in the west. Prices run higher in cities and in antique stores. Prices tend to drop in small towns and flea markets.

Condition

In order to establish a reasonable value for a lantern there must be an understanding of the lantern's condition.

The value of lanterns vary greatly due to condition and authenticity. The condition of paint, tin, nickel, silver, or gold plating affects the price greatly. Solid brass, brass parts, or copper parts add to the value of the lantern even if it is not polished. Good original paint may add slightly to the value. A well executed repaint does not reduce the value greatly as long as it is not hiding rusted, pitted, or damaged metal. Soot, dust, and dirt does not affect the value as lanterns can be easily cleaned.

The following table creates a rating of lantern condition for use by buyers and sellers. The lantern that appears desirable at first, may not be so desirable when the true condition is analyzed. Condition rating "C" is the normal, average, used lantern condition. Condition ratings of "A" or "B" indicates the lantern is in better than typical condition. The ratings "D" and "F" describe a lantern in poor condition. The + (plus) symbol may be added to the letter grade only if the condition approaches the next higher grade, as in "C+." The - (minus) symbol shall never be used as this is the same as the lower grade with a + symbol. In other words the B- is never allowed because it is the same as a C+.

Table 11.1 Lantern Condition Rating

Rating	Lantern Condition Description
A	New or restored to like new condition as originally purchased from a hardware store. No dents, no scratches, no chips around base, no rust, no soot. New wick. No discoloration of finish or burner. No cracks or chips in globe. No trace of dirt or fuel in the fount. Repainted and/or replated to like new condition.
B	Complete and intact. Perfect globe, undented, light tarnish or paint discoloration in original finish. Charred wick, washable soot, or dirt. Discolored burner or crown. Minor refinish needed to make CONDITION A.
C	Repairable rust. Repaint which may obscure rust or repaired rust. Rust on replaceable parts like the fill cap, burner, or globe plate. Single or multiple small dents that do not detract from the look or operation. Easily replaceable part(s) damaged or missing (fill cap, globe, or wick). May be restorable to CONDITION A or B.
D	Missing metal or pits due to rust. Weeping fount. Any hole. Major dent. Previous repairs. Major part missing or damaged (cracked globe, missing burner, rusty crown or globe plate). Restorable to CONDITION C but not to CONDITION A or B.
F	Parts only. Contains one or more parts usable for repairing another lantern. Major parts missing and metal missing due to heavy surface rust or rust out. Not restorable to CONDITION A or B.

Plate 11.2
This unmarked early No. 0 lantern is not found in any price guide yet its approximate value can be calculated at $144 to $270 U.S.D

Values

Now that the condition and age is established, use the Lantern Value Calculator (LVC) in Table 11.2 to score the lantern on its desirable features. Don't skip any boxes. Remember a pre 1895 lantern also gets points for being a pre 1911 and a pre 1935 design.

Take the 1875 vintage lantern of Plate 11.2 as an example. The lantern is in condition C+ (complete and correct parts, no rust, repainted, repairable hole in burner chamber). It has a small fill cap and solder fount construction that places its design before 1895. Score 8 for pre 1895, score 4 more for pre 1911, and score 2 more for pre 1935. Score 0 for the tin fount, 0 for no brass, 2 for the small company, 0 for the lack of markings. Score 0 for the correct but clear globe. Score 0 for burner size and globe size but score 1 because it was made outside of New York. The lantern is a general purpose lantern so score 0 for special purpose and score 1 because the lantern is complete. The total lantern score is 18.

The lantern score determined from the Lantern Value Calculator should be about the same regardless of who does the analysis. The more difficult calculation is putting a value to the score. Using the LVC, a complete, working, used 1936 Dietz Monarch in average condition scores 2. Let's say the 1936, Monarch readily sells for $20.00 in your area. This makes the value of late tubular lanterns about $10 per point. The value per point will vary by regional demand. If you're in a hot market, use $11 to $15 (U.S. dollars) per point. In a cool market, try $8 to $9 per point.

Table 11.2 Lantern Value Calculator

Description	Score
Pre 1869 design or hand made lantern.	*score 18 for condition rating B* *score 10 for condition rating C* *score 6 for condition rating D*
All pre-1895 design or hand made tubular or dead flame lantern.	*score 12 for condition rating B* *score 8 for condition rating C* *score 4 for condition rating D*
All pre-1911 design tubular, or dead flame lantern.	*score 8 for condition rating B* *score 4 for condition rating C* *score 2 for condition rating D*
All pre-1935 design lantern.	*score 3 for condition rating B* *score 2 for condition rating C* *score 1 for condition rating D*
Brass, copper, non-ferrous or glass fount.	*score 5 for condition rating B* *score 4 for condition rating C* *score 3 for condition rating D*
Lanterns made all of brass.	*score 6 for condition rating B* *score 5 for condition rating C* *score 4 for condition rating D*
Not Dietz, Embury, Adams & Westlake (Adlake).	*score 2*
Owners name embossed on globe or lantern.	*score 3 for any railroad compan* *score 2 for any municipal service* *score 1 for any distributor*
Marked for any narrow gauge railroad.	*score 3*
Correct and undamaged globe.	*score 0 for clear* *score 1 for red/ruby* *or score 10 for any other color*
Globe with manufacturers' mark	*score 1*
Number 2 or larger globe.	*score 2*
Number 3 or larger burner.	*score 2*
Made in Canada or the U.S.	*score 1*
Special purpose, rare, boat, bicycle, auto, fire, mill, wagon, dash, RR style, buggy, inspector, beacon, side, street, bull's eye, etc.	*score 2 each*
Complete and functional.	*score 1*
Multiply score by $8 to $15 per point. TOTAL VALUE:	

Lantern Value Table

Rather than do the calculation yourself, Table 11.3 lists the U.S. dollar value for many common lanterns. The condition rating analysis in Table 11.1 must be used to evaluate a lantern's condition but the Lantern Value Calculator (Table 11.2) needs to only be used when the lantern is not listed in the following table.

This table gives the estimated, high, typical, and low values using the Lantern Value Calculator and an average market estimate of $10 U.S. per point.

Only condition ratings B, C and D are included because these are the most common, and because the sky is the limit for any antique in "like new" condition (condition rating A). The auction prices that make news (i.e. forty thousand dollars for a single comic book, etc.) are due to the extreme rarity of a perfect antique. The same holds true for lanterns. The table assumes the lantern has a clear or ruby globe.

Table 11.3, Lantern Value Table in U.S. ($) Dollars (sorted by company)

	MAKER/NAME	CONDITION				MAKER/NAME	CONDITION		
		D	C	B			D	C	B
1.	Adlake Classification	200	270	360	43.	Dietz Junior (early tin)	30	40	50
2.	Adlake Kero	50	60	70	44.	Dietz Junior (late brass)	60	80	100
3.	Adlake Marker	200	270	360	45.	Dietz Junior (late tin)	30	40	50
4.	Adlake Reliable	90	100	110	46.	Dietz Junior (old brass)	60	80	100
5.	Adlake Semaphore	80	90	100	47.	Dietz Junior Wagon	50	60	70
6.	Adlake Switch	200	270	360	48.	Dietz King	130	170	230
7.	Adlake Tail	110	120	130	49.	Dietz Little Giant	10	20	30
8.	Angemeldet	20	20	20	50.	Dietz Little Star	90	160	250
9.	Buhl (brass fount)	80	100	120	51.	Dietz Little Wizard ('38)	20	20	20
10.	Buhl (tin)	50	60	70	52.	Dietz Little Wizard No. 1	20	20	20
11.	Chalwyn Tropic	10	10	10	53.	Dietz Monarch #10	20	20	20
12.	Defiance Dash	60	70	80	54.	Dietz Monarch 1905	20	30	40
13.	Defiance No. 200	50	60	70	55.	Dietz Monarch 1913	20	30	40
14.	Defiance Perfect	50	60	70	56.	Dietz Monarch 1936	10	10	20
15.	Defiance Searchlight	60	80	100	57.	Dietz New Farm	150	260	430
16.	Dietz #2 Wizard (brass)	140	160	180	58.	Dietz Night Watch	20	20	20
17.	Dietz #2 Wizard (tin)	50	60	70	59.	Dietz No. 6 RR Style	60	70	80
18.	Dietz ACME	50	60	70	60.	Dietz No. 15 Wall	70	100	150
19.	Dietz Air Pilot No. 8	20	20	20	61.	Dietz No. 25 Wall	70	100	150
20.	Dietz Beacon No. 30	50	60	70	62.	Dietz No. 39 Standard	60	70	80
21.	Dietz Beacon No. 60	100	150	200	63.	Dietz No. 39 Vulcan	60	70	80
22.	Dietz Blizzard #80	20	20	20	64.	Dietz No. 100	10	20	30
23.	Dietz Blizzard ('12) brass	60	80	100	65.	Dietz O.K.	90	160	250
24.	Dietz Blizzard ('12) tin	30	40	50	66.	Dietz Pioneer Hanging	150	220	310
25.	Dietz Blizzard ('39)	20	20	20	67.	Dietz Pioneer Side	150	220	310
26.	Dietz Blizzard No. 1	90	160	250	68.	Dietz Pioneer Street	150	220	310
27.	Dietz Buckeye Dash	70	80	90	69.	Dietz Roadster Wagon	50	60	70
28.	Dietz Champion	100	140	200	70.	Dietz Royal	30	40	50
29.	Dietz Comet or #50	10	20	20	71.	Dietz Scout	70	100	150
30.	Dietz Crescent	30	40	50	72.	Dietz Standard Deck	130	170	230
31.	Dietz Crystal	120	200	260	73.	Dietz Steel Clad	60	70	80
32.	Dietz D-Lite ('13) brass	80	100	120	74.	Dietz The Original	10	20	20
33.	Dietz D-Lite ('13) tin	50	60	70	75.	Dietz Torch	20	20	20
34.	Dietz D-Lite ('19)	30	40	50	76.	Dietz Traffic Gard	10	20	20
35.	Dietz D-Lite ('39)	10	20	20	77.	Dietz U.S.	90	160	250
36.	Dietz D-Lite No. 90	10	20	20	78.	Dietz Vesta (early brass)	100	190	260
37.	Dietz Eureka	70	100	150	79.	Dietz Vesta (early tin)	70	100	150
38.	Dietz Hy-Lo	10	20	30	80.	Dietz Vesta (short brass)	120	110	180
39.	Dietz Ideal	50	60	70	81.	Dietz Vesta (short tin)	50	60	70
40.	Dietz Iron Clad	90	160	250	82.	Dietz Victor	20	30	40
41.	Dietz Junior #20 (brass)	50	60	70	83.	Dietz Wizard ('00)	20	20	20
42.	Dietz Junior #20 (tin)	20	20	20	84.	Embury Camlox	60	80	100

	MAKER/NAME	CONDITION		
		D	C	B
85.	Embury Little Air Pilot	20	30	40
86.	Embury Little Supreme	20	30	40
87.	Embury Luck-E-Light	20	25	30
88.	Embury No. 0 Air Pilot	20	30	40
89.	Embury No. 2 Air Pilot	20	30	40
90.	Embury Supreme	20	30	40
91.	Embury Traffic Gard	20	20	20
92.	Feuerhand Baby	10	10	10
93.	Feuerhand Nier	30	30	30
94.	Fixed Globe Oil Lanterns	190	300	470
95.	Ford Side Lamp	90	120	170
96.	Ford Tail Lamp	90	120	170
97.	H.S.B. & Co. Bantie	200	290	400
98.	Ham Gem (brass)	200	290	400
99.	Ham Gem (tin)	110	180	270
100.	Ham No. 2 (CB)	40	50	60
101.	Ham No. 2 (HB)	40	50	60
102.	Ham Nu-Style (tin)	70	80	90
103.	Ham Nu-Style (brass)	100	120	140
104.	Ham's Clipper 1893	30	40	50
105.	Ham's Clipper 1899	30	40	50
106.	Handlan "The Handlan"	70	80	90
107.	Handlan St. Louis	50	50	50

	MAKER/NAME	CONDITION		
		D	C	B
108.	Hurwood Aladdin	70	100	150
109.	Kwang Hwa	10	10	10
110.	Nail City Crank	130	200	290
111.	Perkins Perko	160	210	280
112.	Prisco No 477	50	60	70
113.	Prisco No 321	70	80	90
114.	Prisco No. 331	70	80	90
115.	Rayo No. 82	80	100	120
116.	Rayo Standard (brass)	200	290	400
117.	S.G. & L. Buckeye	200	290	400
118.	S.G. & L. Co. No. 0	110	180	270
119.	S.G. & L. L W	110	180	270
120.	Sherwoods Ltd.	40	50	60
121.	Trulite	50	60	70
122.	Universal Deck (brass)	220	310	420
123.	W.M. Co. Standard	130	200	290
124.	Warren STA-LIT	50	60	70
125.	Wheeling Leader	50	60	70
126.	Wheeling Paull's	50	60	70
127.	Wheeling Regal	50	60	70
128.	Winged Wheel 350	10	10	10
129.	Winged Wheel 400	10	10	10
130.	Winged Wheel 2000	10	10	10

53054...$90.00

FRENCH HURRICANE LAMP

F. Manufactured by the respected Guillouard company for over 80 years, this hurricane lamp is beautifully made for decades of service. Famous for staying lit in severe weather, Guillouard lamps are the first choice of sailors around the world. During mosquito season, you might fill yours with citronella oil and enjoy it on the front porch or deck. Choose Original Tin or luxurious Solid Brass. Imported from France. 12"h x 6" diameter.

Original Tin • 54887...$45.00
Solid Brass • 54885...$180.00
16 oz. Lamp Oil • 22818...$5.00

F

Plate 11.3
This ad gives some idea of the price difference between a tin lantern and the same lantern in solid brass.

Plate 11.4
Two Adlake Reliables in fresh paint. One is a collector's item and the other is almost worthless. The lantern on the left was electrified and contains no usable parts.

Chapter 12

Tips on Collecting

Rarity

1. Lanterns made of brass or those with brass or copper parts are always more desirable. Carry a magnet to check for non-ferrous founts and tubes.
2. Always carry this and other reference books.
3. Blue, yellow, green and multi- color globes are more desirable than clear and ruby.
4. Look for the older style pear shaped globes.
5. While you're looking at the globe, see if it has a bull's eye lens.
6. A small, one piece brass fount cap is a good indication of a pre World War 1 lantern.
7. Look for square, round, and soldered tubes. The less machine forming, the better.
8. The more hand soldering in a lantern's construction the older it is.
9. Dietz, Embury, Defiance, and Ham account for the majority of all lanterns produced. Watch for other smaller manufacturer's products.
10. Don't overlook new lanterns. Some are reproductions using old dies but all are unique and produced in limited quantities.

Condition

1. When a lantern is found, look at the condition of the bottom first. If it is clearly undamaged and unrusted then the lantern will be easier to make functional.
2. Check the bottom of the tubes for rust holes. Dirt and soot collect here and hold moisture. Tube holes are difficult to repair.
3. Next check the globe for cracks as old globes in good condition are difficult to find.
4. Does the lantern have the correct burner? If the lantern has a small fill cap then the burner should likely be made of brass. For lanterns with large fount caps, steel burners are the norm.
5. Check for dents in the fount and tubes where access from behind is not possible. These are difficult to repair. Dents in the crown and globe plate can be pressed out.
6. Now look at the finish. Don't reject a lantern based on paint condition or rust alone. These problems can be fixed. Some think a cracked and flaky finish adds to the lantern's charm.

Record Keeping

1. Use a ruler to help identify the exact model lantern.
2. Even if you decide not to buy, note the lantern make, model, condition and price for future reference.

Services and Information

1. Eastwood Restoration Supplies
1-800-343-9353
www.eastwoodco.com
Materials and supplies for antique refinishing including paint stripper, rust remover, primer, paint, gas tank sealer, plating solution, and buffing supplies.

2. Key, Lock and Lantern
www.klnl.org
Railroadiana collector organization with quarterly newsletter.

Web Directory

1. There is a discussion group page that is mostly for traffic lanterns but, all kerosene lanterns are tolerated: http://members3.boardhost.com/trafficgard/
2. A comprehensive web site with parts and lanterns for sale: http://www.lanternet.com
3. Great site about traffic lanterns: http://home.earthlink.net/~trafficgard/
4. Schiffer collector books http://www.schifferbooks.com
5. The author's kerosene lantern pages: http://www.classiclantern.com

Plate 12.1
This is a close-up view of the Automatic Extinguisher on the Dietz Pionveer Street Lamp of Plate 6.3. The large knob holds the wick out of the oil after a measured number of hours have passed. This timer is marked 10, 15, 20, and 25 hours. Note the match striker to the right of the timer.

Plate 12.2
The Dietz Tubular Square Station Lamp (1880-1887) also uses a self-extinguisher.
To learn more about this lantern visit: www.classiclantern.com

References

Barrett, Richard C. *The Illustrated Encyclopedia of Railroad Lighting, Volume 1, The Railroad Lantern;* Rochester, New York: Railroad Research Publications

Barrett, Richard C. *The Illustrated Encyclopedia of Railroad Lighting, Volume 2, The Railroad Signal Lamp;* Rochester New York: Railroad Research Publications.

Beitz, Les. *Treasury of Frontier Relics:* Conroe, Texas: True Treasure Library, 1971

Blumenstein, Lynn. *Treasure Hunters Relic Identification:* Salem, Oregon: Old Time Bottle, 1968

Dreimiller, David. *The Dressel Railway Lamp & Signal Co.:* Hiram Ohio: Hiram Press, 1995

Hobson, Anthony. *Lanterns That Lit Our World, Book One.* Spencertown, New York: Golden Hill Press, Inc., 1991

Hobson, Anthony. *Lanterns That Lit Our World, Book Two.* Spencertown, New York: Golden Hills Press, Inc., 1997

Lewerenz, Alfred S. *Antique Auto Body Accessories for the Restorer.* Arcadia, California: Post Motor Books, 1970

L-W Book Sales. *Collectible Lanterns.* Gas City, Indiana: L-W Book Sales, 1997

Mehlman, Felice. *Phaidon Guide to Glass.* London, England: Prentice-Hall, Inc., 1983

Sears, Roebuck and Co. Catalogue, Chicago, Illinois: Sears, Roebuck Co., 1902

Wood, Neil S. *Collectible Dietz Lanterns.* Gas City, Indiana: L-W Book Sales, 1997

Glossary

ball usually arched, wire, wood or metal handle used to carry a lantern
bayonet fount fount that twists off, usually from below
bell crown, cap above the globe on a hot blast lantern
bull's eye magnifying lens attached to globe plate or formed as part of the globe
burner assembly that includes the wick and wick adjustment mechanism
candle wax or tallow containing a wick that may be burned for light, heat, or votive purposes
capillary action action by which molecules of liquid travel through porous materials
chimney globe or tube above flame used to create a draft
cold blast or C.B. refers to lanterns which draws fresh air in to the draft chamber
crizzling diseased glass, degradation of glass due to inaccurate mixing of components
crown cap at the top of hot and cold blast lanterns
D-Lite trade name given to a series of lanterns using the large, short globe
dead flame refers to lanterns having no tubes to return hot air to the draft chamber
draft chamber area below the flame where one or more draft sources collect
draft inlet opening, hole or screen to allow fresh air to enter the burner
drop fount held in place by side springs and removed from the bottom
fill hole in fount used to add fuel
fill cap plugs fill hole to prevent fuel spill
flame source of illumination, burning fuel
fluted pear shape, shape of early globes
fount reservoir for fuel, usually of metal or glass
fuel kerosene, sperm whale oil, coal oil, paraffin, etc.
globe glass enclosure for flame that protects flame from wind, usually clear or colored glass
globe guard usually part of the perforated plate, metal wires to protect the globe from breakage
globe plate perforated plate, metal disk, usually full of holes, at the bottom of the globe
globe, short large squat globe invented in 1912 for large kerosene burners
handle fixed grip of metal or wood, see also bail
hot blast refers to lanterns using exhaust gas to shape the flame
hydrocarbon an organic compound containing hydrogen and carbon atoms often occurring in petroleum, natural gas, and coal
insert fount installed from the top by removing the globe
Junior name given to line of small lanterns
kerosene petroleum-derived, light oil used as jet fuel and lamp and lantern fuel
knob wick adjust, used to regulate the wick height
LOC-NOB lock knob. Ears on a globe used to hold it in the guard
No. 0 common globe size
perforated plate globe plate, metal disk, usually full of holes, at the bottom of the globe
pot fount, usually used when referring to railroad lanterns
rising cone burner where the dome is attached to the globe plate and lifts to allow access to the wick
short globe globe for a lantern made to burn kerosene, less than 5 inches tall, e.g. vesta, kero
side lift handle, or mechanism used to lift the globe for access to wick or flame
soot also called carbon black an almost pure carbon that collects on globe and chimney
square lift finger grip at lantern crown or bell used to lift the globe
tall globe globe for a lantern made to burn sperm, whale, signal, or lard oil, usually 5.38 to 6 inches tall
tubular any lantern type with one, two or more air tubes
vapor/vaporized gaseous state
volatile readily vaporized
Vulcan line of railroad conductor's or brakemen's lanterns
warm air tube hollow tube(s) used by tubular lanterns to create a draft by carrying warm air to the flame
wick absorbent material used to carry fuel to flame, typically cotton or felt
wick adjust knob or wire loop used to adjust the wick up or down
wizard lantern name and name of a small globe

Back cover: The Dietz No. 30 Beacon (left) described in Plate 9.12 and it's big brother, the No. 60 Beacon. The large Beacon uses a Number 3 burner and rare Number 2 globe.

Index